DNA TECHNOLOGY

Selected Titles in ABC-CLIO's
**CONTEMPORARY
WORLD ISSUES**
Series

For a complete list of titles in this series, please visit
www.abc-clio.com.

Books in the Contemporary World Issues series address vital issues in today's society, such as genetic engineering, pollution, and biodiversity. Written by professional writers, scholars, and nonacademic experts, these books are authoritative, clearly written, up-to-date, and objective. They provide a good starting point for research by high school and college students, scholars, and general readers as well as by legislators, businesspeople, activists, and others.

Each book, carefully organized and easy to use, contains an overview of the subject, a detailed chronology, biographical sketches, facts and data and/or documents and other primary-source material, a directory of organizations and agencies, annotated lists of print and nonprint resources, and an index.

Readers of books in the Contemporary World Issues series will find the information they need to have a better understanding of the social, political, environmental, and economic issues facing the world today.

DNA TECHNOLOGY

A Reference Handbook

David E. Newton

**CONTEMPORARY
WORLD ISSUES**

A B C ⬥ C L I O

Santa Barbara, California • Denver, Colorado • Oxford, England

Library of Congress Cataloging-in-Publication Data

Newton, David E.
   DNA technology : a reference handbook / David E. Newton.
      p. cm. — (Contemporary world issues)
   Includes bibliographical references and index.
   ISBN 978–1–59884–328–6 (hard copy : alk. paper)—ISBN 978–1–59884–329–3 (ebook)
1. Genetic engineering—Handbooks, manuals, etc. I. Title.   [DNLM:
1. DNA. 2. Genetic Techniques. 3. Biotechnology. 4. Genetics. QU 450 N562d 2009]
QH442.N465   2010
660.6′5—dc22   2009036041

14 13 12 11   2 3 4 5

This book is also available on the World Wide Web as an eBook.
Visit www.abc-clio.com for details.

ABC-CLIO, LLC
130 Cremona Drive, P.O. Box 1911
Santa Barbara, California 93116-1911

This book is printed on acid-free paper ∞

Manufactured in the United States of America

This book is dedicated to
John Fellows, Jay Moore, and Larry Thon
Good travelers and good friends

Contents

Preface

On April 25, 1953, American biologist James Watson and English chemist Francis Crick published a paper in the British journal *Nature* proposing a chemical structure for a family of chemical compounds known as deoxyribonucleic acid (DNA). The structure they suggested consisted of a long pair of chains twisted around each other, somewhat like a spiral staircase, with subunits called nitrogen bases paired with each other as steps on the staircase. At the conclusion of their paper, Watson and Crick noted that, "It has not escaped our notice that the specific pairing we have postulated immediately suggests a possible copying mechanism for the genetic material." What Watson and Crick had discovered is that DNA molecules carry a "code" for genetic characteristics, such as hair and skin color, ear shape, color of eyes, height, and a host of other traits. This discovery solved a problem that had troubled scientists for decades: How does like produce like during the process of reproduction? How does a willow tree always produce other willow trees only, and never oak trees? How do redheaded children appear only in families with redheaded parents, grandparents, and/or great-grandparents, and never in families whose predecessors had only black or brown hair?

The very remarkable thing about the Watson and Crick discovery is that the answer to these and other biological questions was given in chemical terms. After Watson's and Crick's paper, scientists were able to think about how hereditary traits are transmitted in terms of atoms and molecules. And, since chemists already knew a great deal about manipulating atoms and molecules, the possibility presented itself that biologists might now be able to manipulate the characteristics of living organisms in much the same way: by moving around atoms and molecules.

Today, because of this discovery, biological studies are as much a function of chemical changes as are industrial processes, such as the production of paper, the manufacture of plastics, and the processing of metallic materials.

An even more profound consequence of the Watson and Crick discovery is that one can now say that, to a very considerable extent, it is possible to describe everything we know about the biological characteristics of a human being (or a chimpanzee or elephant or flea) by writing chemical formulas and chemical equations. To the average person, the expression below may just be a series of letters, or, to someone with an introductory course in chemistry, a series of nitrogen bases in a DNA molecule. But to a molecular biologist, that string of letters stands for some physical trait, such as hair color or a potential health problem, such as sickle-cell anemia.

```
-T-T-T-C-A-C-G-G-G-C-C-A-T-T-A-C-G-G-C-G-C-T-T-A-T-
-A-A-A-G-A-G-C-C-C-G-G-T-A-A-T-G-C-C-G-C-G-A-A-T-A-
```

This possibility was apparent to scientists from almost the moment that Watson and Crick published their 1953 paper. Researchers began to ask if it might be possible to alter the "genetic code" in an amoeba, or a fruit fly, or even a human being in order to change that organism's genetic composition and, therefore, its physical characteristics. Out of that line of thinking has come a whole new field of study that uses information about DNA structure to develop practical applications. This book reviews those applications in six fields: forensic science, the development of transgenic organisms (organisms with genetic characteristics of two different species), genetic testing, gene therapy, molecular farming (also known as *pharming*), and cloning.

Perhaps the most remarkable feature of the history of DNA technology is the tempo at which progress has been made. Only 50 years after the Watson and Crick discovery, scientists are applying that knowledge to identifying genetic diseases in fetuses, altering tobacco plants so that they will produce useful drugs, identifying criminals accused of crimes with a certainty of one in a billion or better, treating (and, perhaps, curing) adults with life-threatening genetic disorders, and producing multiple copies (clones) of animals born outside the normal process of reproduction.

Chapter 1 provides a historical review of the development of these six DNA technologies, with an overview of the scientific principles fundamental to each technology. Chapter 2 reviews some social, ethical, political, and other issues that have arisen as a result of the development of these technologies. In many cases, the use of DNA technology has raised some fundamental philosophical and theological questions about what it means to be human, or what it means to be alive. One should not be surprised that these questions engender serious, and often heated, debates as to the conditions under which a particular technology should be used or, indeed, if it should ever be used at all. Chapter 3 focuses on the international aspects of the debates over DNA technology. Perhaps the most significant findings of this chapter are the enormous diversity in the use of DNA technologies in various parts of the world, the extent to which these technologies have stirred up controversy over their use, and the range of views that people from different parts of the world have about DNA technology.

Chapters 4 through 8 provide a variety of resources to help the reader continue a study of DNA technology. Chapter 4 offers a chronology of important events in the growth of scientific knowledge about the nature of DNA and its role in biological processes, along with some important political, social, legal, and other events related to this growth of knowledge and its applications in everyday life. Chapter 5 provides brief biological sketches of some important figures in the development of DNA technology and its applications. Chapter 6 contains some political and legal documents that illustrate the way the United States and other countries around the world are attempting to assess, regulate, and control the applications of DNA technology in a variety of fields. Some relevant data about the growth of this field of study are also included. Chapter 7 provides a list of organizations interested in DNA technology, some whose primary responsibility is regulatory, some interested in advancing the development and use of DNA technology, and some working to moderate or stop the development of that field of research. Chapter 8 is an annotated bibliography of print and electronic resources for the six major types of DNA technology discussed in the book. The book concludes with a glossary of important terms used in the field of DNA technology.

1

Background and History

Introduction

Are you in the market for a new pet? How about a fish that glows in the dark? Even under white light, this new GloFish® is a striking shade of red, green, or yellow; you get to choose from Starfire Red®, Electric Green®, or Sunburst Orange®. GloFish® are one of the somewhat unexpected products of DNA technology. They were first invented in 1999 by scientists at the National University of Singapore who were trying to develop a fish that could be used to warn of the presence of pollutants in water. They added a gene that codes for a protein known as green fluorescent protein (GFP) that occurs naturally in jellyfish. Their plan was to further program the fish to display its color when a particular pollutant was present in the water. While the research was still underway, entrepreneurs from a U.S. company, Yorktown Technologies, L.P., met with inventors of the GloFish® to work out rights for licensing and selling the engineered fish in the United States. After receiving approval from all relevant U.S. regulatory agencies, Yorktown Technologies began selling GloFish® in late 2003. The age of engineered life forms has arrived in your living room!

Modifying Life: The Early History

Humans around the world today have a host of plants and animals at their disposal as a source of food, to help in daily work, to serve as helping animals and pets, and for countless other purposes. They range from strawberries and corn and cabbage to horses and dogs and sheep. All of these organisms have evolved over long periods of time from one or more primitive forms. Sometimes evolution has occurred naturally, without any efforts on the part of humans. For example, two closely related species may mate by one method or another and a new, slightly different species is produced. On other occasions, humans have attempted to breed organisms with each other in order to produce plants and animals that are healthier, better tasting, more attractive, stronger, or better in some other way than their progenitors. These efforts resulted in the domestication of many plants and animals long before the period of written history: by 13000 BCE, in the case of rice; by 10000 BCE in the case of dogs, goats, pigs, and sheep; by 7000 BCE in the case of the potato; and by 2700 BCE in the case of beans, chilies, corn, and squash (Macrohistory and World Report 2009). Evidence for the hand pollination of plants—date palms—dates to the Assyrians and Babylonians in about 700 BCE (History of Plant Breeding 2009).

In many cases, humans controlled the reproduction of plants and animals without any knowledge of the scientific principles involved; they simply applied information gained from centuries of experience based on trial-and-error experiments. They bred sheep with the best wool, potatoes with the best taste, horses with the greatest strength, corn that was most nutritious, and other plants and animals with desirable properties. Without understanding how, they directed the evolutionary paths of untold numbers of plant and animal species.

In some cases, they even produced completely new plants and animals by hybridization (the crossbreeding of different species or varieties of plants or animals). One of the earliest examples may have been the mule and the hinny. A mule is the offspring of a male donkey (a jackass) and a female horse (a mare), while a hinny is the offspring of a female donkey (jenny) and a male horse (a stallion). Archaeologists have found representations of mules and hinnies in Egyptian tomb paintings dating back to at least 1400 BCE (Sherman 2002, 42). Throughout the centuries, humans have produced a number of other hybrid animals in much the same way the mule and hinny must first have appeared: the beefalo (a cross between an American bison and a domestic cow), Bengal cat (domestic cat and Asian leopard cat), zeedonk (donkey and zebra), dzo (domestic cow and yak),

Leyland cypress (Monterey cypress and Nootka cypress), peppermint (water mint and spearmint), and loganberry (raspberry and blackberry). Although most of the details of the ways in which humans have influenced the evolution of plants and animals by hybridization are unavailable (because they occurred so long ago), it is clear that that influence has been profound: plants and animals in the twenty-first century are very much the product of human modification.

The Birth of Genetics

By the last quarter of the seventeenth century, hybridization had taken on a new, more scientific, flavor, with breeders and farmers using a more rational approach to the crossing of species. Horticulturists in The Netherlands were a particularly good example, having found dependable methods for producing a large variety of new plants, including the first hybrid hyacinth and a number of new tulip varieties (Murphy 2007, 246, 331). Without much doubt, however, the first true research designed to produce a scientific understanding of the process by which genetic traits are passed from one generation to the next was conducted by the Austrian monk, Gregor Mendel, between 1856 and 1863.

Mendel brought to his research a valuable combination of interests in the evolution of organisms, natural history, and the physical sciences. He studied seven physical properties of the common pea plant (*Pisum sativum*): flower color (purple or white), flower position (axial or terminal), stem length (short or long), seed shape (round or wrinkled), seed color (yellow or green), pod shape (constricted or inflated), and pod color (yellow or green). His experiments were relatively straightforward. Pea plants with various combinations of these traits were crossbred with each other, and the characteristics of offspring were noted. The offspring were then crossbred with each other, and characteristics of the next generation were also recorded. In this way, the transmission of traits from parents to the first generation to the second generation, and so on, could be observed. Mendel found, as one simple example, that the cross between a pea plant with yellow seeds and a plant with green seeds produces offspring all of whom have yellow seeds only. If the offspring (the f1 generation) are then crossbred among themselves, plants with green seeds reappear in one out of four cases in the next (f2) generation. Further, the 3:1 ratio of traits continues to reappear in future (f3, f4, etc.) generations (For an excellent overview of Mendel's work, see Mawer 2006).

Mendel's work is significant for a number of reasons. First, it contradicted the current belief that genetic traits tend to blend in the passage of generations. That is, classic theory at the time would have predicted that the cross described above would have resulted in pea plants with greenish-yellow seeds in some later generation, with the shade of green-yellow varying from plant to plant. Second, it suggested that the transmission of hereditary characteristics requires the presence of some kind of "carrying unit" for a genetic trait. Mendel referred to these units as *factors*, *unit factors*, or, simply, *units*, but he really had no idea as to what these units were. Third, Mendel demonstrated that an organism might possess the unit factor for a trait, but might not actually show that trait. In the example above, for example, pea plants in the f1 generation all had yellow seeds. But they must somehow have retained the unit factor for green color, because in the next generation one out of four plants had green seeds. Based on these observations, Mendel also concluded that each offspring plant must receive one unit factor from each parent. Finally, he hypothesized that, for each trait, one unit factor must be dominant, and the other recessive. That is, plants that inherited both yellow and green unit factors actually produced only yellow seeds, so the yellow factor must be dominant over the green factor. Only plants that received two green unit factors from parents could produce green seeds.

Mendel reported the results of his research at two meetings of the Natural History Society of Brno (also Brünn) on February 8 and March 8, 1865. He then published the text of his remarks in a paper, "Versuche über Pflanzen-Hybriden" ("Experiments on Plant Hybridization"), in the *Proceedings of the Natural History Society of Brno for the year 1865* (published in 1866). His verbal and written reports were almost entirely ignored. Over the next 35 years, his work was cited only a handful of times, and it had essentially no effect on biological thought. Only in 1900 did Mendelian genetics once more see the light of day when three biologists (Hugo de Vries, Carl Correns, and Erich von Tschermak), all working independently of each other, rediscovered Mendel's work and brought that work to the attention of the biological community. Mendel had simply been born three decades too early!

For more than a century, Mendelian genetics has been a powerful tool for understanding the transmission of hereditary traits. Geneticists now deal with much more complex systems than those used by Mendel, but the principles he elucidated—modified and updated—continue to explain most of the basic facts of inherited traits. The one important problem that remained unsolved for decades after Mendel's rediscovery in 1900 was the nature of the transmitting unit, the unit factor that Mendel invented to explain his findings. Later biologists sometimes invented new names for the unit factor. Hugo DeVries himself suggested the term *pangen* in 1889, long before he knew of Mendel's work, and the modern name for the unit factor—*gene* —was invented in 1903 by Danish botanist and geneticist Wilhelm Johannsen. The fundamental problem was that no one really had any idea as to what a unit factor (or pangen or gene) was. The naming game was simply a method for identifying a "black box" in organisms responsible for transmitting genetic traits.

In 1953, that problem was finally resolved. American biologist James D. Watson and British chemist Francis Crick discovered that the "black box" is a chemical compound, deoxyribonucleic acid, or DNA.

The Road to DNA

Like virtually all scientific breakthroughs, the discovery of DNA did not spring fully formed from Watson's and Crick's brains. It was the result of decades of research and theorizing about the nature of the genetic material. DNA had been discovered as far back as 1869 by Swiss physician and biologist Johannes Friedrich Miescher, although he knew of no biological role for the compound (which he called *nuclein*). In fact, scientists knew very little about the structure or function of DNA for nearly seven decades after its discovery. For most of that time, biologists were convinced that unit factors (or genes) must consist of some kind of protein-like material. Proteins are very large, complex molecules consisting of long chains of amino acids. By "very large," one means molecules contain tens of thousands, hundreds of thousands, or millions of atoms. It made sense to believe that molecules of such complexity would be the ideal carriers of complex genetic information. By contrast, DNA is a relatively simple molecule consisting of only three basic units: a sugar (deoxyribose), a phosphate group, and one of four nitrogen bases.

One of the first clues in the DNA story came in 1909, when Russian-American chemist Phoebus Levene discovered that one type of nucleic acid contains a sugar called *ribose*. Levene called that type of nucleic acid *ribonucleic acid*, or RNA. Twenty years later, Levene found a second sugar in other types of nucleic acid, the sugar deoxyribose. Deoxyribose differs from ribose only in that it contains one less hydroxyl group (-OH); thus, the name, *de-* ("without") *-oxy-* ("oxygen") *ribose*. To these nucleic acids he gave the name *deoxyribonucleic acid*, or DNA. A hint that some form of nucleic acid might be involved in heredity came in 1924 when researchers found that both protein and nucleic acids were present in chromosomes. This information was essential for the identification of genes, of course, because biologists already knew that genes are located on chromosomes. This discovery meant that both proteins and nucleic acids were candidates for the role of genes. It did not tell which family of compounds made up genes, and scientists continued to favor a protein hypothesis.

Another clue to the puzzle came in 1928 when British medical officer Franklin Griffith found that the hereditary material (genes) was something that was stable when heated, a characteristic not common with proteins. Griffith killed bacteria by heating them and then inserted their cellular components into live bacteria. He found that genetic information from the killed bacteria was expressed within the living bacteria. The answer as to which heat-resistant compound it was that produced this result was answered in 1944 in a series of elegant and relatively simple experiments conducted by Oswald Avery, Colin MacLeod, and Maclyn McCarty. The three researchers worked with two types of pneumococci bacteria (the bacteria that cause pneumonia), identified as the R-type (for "rough" coating) and S-type (for "smooth" coating). They removed from the nuclei of S-type bacteria all protein, DNA, and other compounds and transferred each type of compound to R-type bacteria. Of the host bacteria, only those that received DNA were transformed into S-type bacteria. The results were strongly suggestive that DNA was the hereditary material, that is, that genes consisted of DNA in some form. Strangely enough, most researchers were reluctant to accept the implications of the Avery-MacLeod-McCarty experiment (which may explain in part why none of the three ever received a Nobel Prize).

One more piece in the DNA puzzle was put into place in 1949 when Austrian-American Erwin Chargaff discovered that the amount of adenine in a DNA molecule is always the same as the amount of thymine, and the amount of cytosine, the same as the amount of guanine. Adenine, cytosine, guanine, and thymine are four nitrogen bases that make up DNA molecules. Chargaff realized that his discovery was an important clue in unraveling the structure and significance of DNA, but the fulfillment of that understanding was left to Watson and Crick.

By the early 1950s, most of the clues needed for the deciphering of the DNA structure were in place. The key figures in the research were almost within shouting distance of each other, Watson and Crick at the Cavendish Laboratory at the University of Cambridge, and Maurice Wilkins and Rosalind Franklin at King's College, London. While Watson and Crick were manipulating the information they had about DNA from Avery, Chagraff, and other researchers, Wilkins and Franklin were taking X-ray photographs of the DNA molecules. X-ray crystallography is a very powerful, but very difficult, technique for determining the structure of complex molecules by shining X rays through them. The images obtained by this method are often very difficult to interpret and to translate into three-dimensional models of the molecules. By late 1952, Franklin had obtained extraordinary images of DNA molecules, which appeared to provide the final clue to the compound's structure. She hesitated in announcing her results, however, wanting to further confirm them. Crick and Watson, who had seen her results, were not so shy. They realized that Franklin's images gave them the last bit of information they needed, and by March 7, 1953, had constructed their model of the DNA molecule. They reported the results of their research in a paper that appeared in the April 25, 1953, issue of the journal *Nature*, titled "A Structure for Deoxyribose Nucleic Acid." They began their paper with the comment:

> We wish to suggest a structure for the salt of deoxyribose nucleic acid (D.N.A.). This structure has novel features which are of considerable biological interest. (Watson and Crick 1953, 737–738)

Watson and Crick (with help from Franklin and others) had unraveled the structure of the DNA molecule. The gene had finally been given a clear and unequivocal chemical and physical structure.

The Structure of DNA

So what did Watson and Crick discover? They found that a DNA molecule consists of two very long spaghetti-like strands (the "backbone" of the molecule) made of alternating sugar (deoxyribose) and phosphate groups (see Figure 1.1).

These two strands are wrapped around each other, a bit like a circular staircase, in a geometric form known as a double helix.

Attached to each sugar molecule is one of four nitrogen bases: adenine, cytosine, guanine, or thymine, as shown in Figure 1.2.

The combination of one sugar (deoxyribose) unit and one nitrogen base, as in S – A, is called a *nuceloside*; the combination of a sugar, base, and phosphate group, as shown in Figure 1.3, is known as a *nucleotide*. To a chemist, then, the long string that makes up a DNA backbone is a *polynucleotide*, many ("poly-") nucleotide units joined to each other.

FIGURE 1.1
Backbone of DNA Molecule

S - P - S - P - S - P - S - P - S - P - S - P - S - P - S - P - S - P - S - P - S - P - S -.

FIGURE 1.2
Nitrogen Bases in DNA

```
S - P - S - P - S - P - S - P - S - P - S - P - S - P - S - P - S - P - S - P - S - P - S -
|       |       |       |       |       |       |       |       |       |       |
C       A       C       T       G       G       G       C       A       T       G       C
```

(The placement of nitrogen bases is random in the molecule.)

FIGURE 1.3
Nucleotide

```
S - A
|
P
```

The two strands are then wrapped around each other in such a way that the nitrogen bases are inside the molecule, with complementary bases adjacent to each other, as shown in Figure 1.4.

In this structure, the nitrogen bases are paired with each other so that every adenine is joined to a thymine, and every cytosine to a guanine. How this structure translates into genetic information is discussed later in this chapter.

FIGURE 1.4
Double-helix Structure of the DNA Molecule

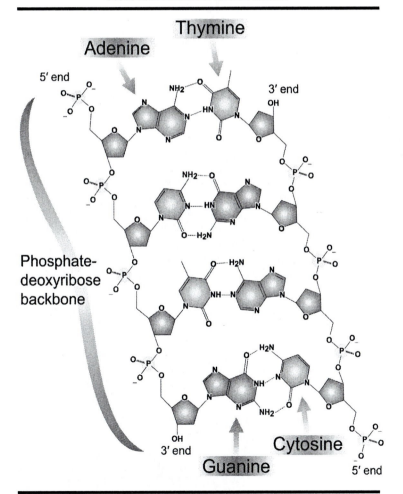

FIGURE 1.5
Structure of an RNA Molecule

```
- P - R - P - R - P - R - P - R - P - R - P - R - P - R - P - R - P - R - P - R - P - R - P - R -
      |       |       |       |       |       |       |       |       |       |       |       |
      A       C       U       G       G       U       C       C       C       G       U       U
```

At this point, a word needs to be said about DNA's perhaps somewhat less well-known cousin, RNA. The two families of compounds are similar in many ways, especially with regard to their importance in the transfer of genetic information from one generation to the next. But structurally, they differ in three important ways. First of all, the sugar in the backbone of an RNA molecule is ribose, not deoxyribose. Second, the four nitrogen bases in RNA are adenine, cytosine, guanine, and uracil, rather than thymine. Third, RNA is a single-stranded molecule and not a double-stranded helix, as in the case of DNA. A typical RNA molecule, then, might have a structure like shown in Figure 1.5.

The Rise of Molecular Genetics

The discovery of DNA structure made by Watson and Crick is sometimes cited as being one of the most important discoveries in the history of science. The reason for this accolade is that prior to the discovery, geneticists knew a great deal about the way genetic traits were transmitted from generation to generation. But they knew nothing at all as to how humans might manipulate that process at the most fundamental level, that of cells and molecules. Beyond trial-and-error hybridization experiments, genetic manipulation was essentially a mystery. After Watson and Crick, scientists realized that the genetic material consisted of molecules—molecules of DNA—that could (at least in theory) be synthesized, broken apart, and manipulated in essentially the same way as any other kind of chemical molecule. The opportunities for engineering genes and altering heredity seemed endless.

Before the "at least in theory" could become a practical reality, however, a great many technical problems for working with DNA molecules needed to be solved. Francis Crick was intimately involved in solving the first two of those problems, the mechanism by which genetic information is coded in a DNA molecule, and the process by which that information is used to synthesize proteins in a cell. In a paper published in 1958, Crick stated two general principles about the function of DNA. The first, which he called the Sequence Hypothesis, says that:

> the specificity of a piece of nucleic acid is expressed solely by the sequence of its bases, and that this sequence is a (simple) code for the amino acid sequence of a particular protein. (Crick 1958, 152)

In other words, the genetic information in a DNA molecule consists of some discrete set of nitrogen bases in the molecule. The sequence ATC might translate into one instruction, the sequence TTC into another instruction, the sequence CGC into yet another instruction, and so on. The most likely—but, again, not known— likelihood was that base sequences carry information about the synthesis of amino acids. Amino acids are the basic units from which proteins are made, and the primary function of DNA is to tell a cell how to make proteins.

At this point, Crick (and all other scientists) knew virtually nothing about the code itself, although one piece of information seemed to be obvious: the number of nitrogen bases needed for a specific instruction. Suppose a single nitrogen base codes (carries the information needed) for the synthesis of some specific amino acid, as, for example: adenine codes for the amino acid aspartic acid. But that code can't work because there are only four nitrogen bases and more than 20 amino acids needed for the synthesis of proteins. Similarly, a two-base code cannot work because the greatest number of amino acids coded for with a two-base system is 16 (4×4). But a three-base code would provide enough combinations to code for all known amino acids found in proteins.

Crick called his second hypothesis the Central Dogma. In its simplest form, this hypothesis says that once "information" has passed into protein it cannot get out again. That is, nucleic acids are able to make other nucleic acids or proteins, but proteins are never able to make nucleic acids.

A number of scientists misunderstood Crick's argument here. They thought that he was saying that the information stored in DNA molecules may be transferred to RNA molecules, which are then used to synthesize proteins. And this information pathway is certainly correct. We now know that it is the primary mechanism by which the information stored in DNA molecules directs the synthesis of proteins. But, as it turns out, other pathways are also possible. For example, RNA molecules are sometimes used to make new DNA molecules. But, as the Central Dogma says, the one prohibited transfer is *from* a protein molecule to any other kind of molecule, protein, or nucleic acid.

The obvious next step in the refinement of DNA technologies was to discover the genetic code, the set of three nitrogen bases (sometimes called a *triad* or a *codon*) that codes for amino acids. That breakthrough occurred in 1966 when American biochemists Marshall Nirenberg and Robert Holley discovered the first element in the code. In an elegantly simple experiment, they prepared a sample of RNA made of only one nitrogen base, uracil. They called the molecule polyuracil RNA because it contained many (*poly-*) uracil units joined to each other. When polyuracil RNA was inserted into a cell whose own DNA had been removed, the cell made only one product, the amino acid phenylalanine. The codon UUU, therefore, codes for the amino acid phenylalanine. The first clue to the genetic code had been produced.

The Nirenberg-Holley experiment not only provided the first "letter" in the genetic code, UUU = phenylalanine, but also suggested a method for breaking the rest of the code. Other researchers were soon constructing synthetic RNA molecules with known base sequences and learning which sequence was responsible for the synthesis of which amino acid. Within a short time, the complete genetic code was known. That code is shown in Figure 1.6.

Recombinant DNA Technology

By the mid-1960s, scientists had developed a reasonably satisfactory model of the way DNA molecules reproduce themselves and direct the synthesis of proteins in cells. They were beginning to imagine ways in which they could manipulate that process artificially. In order to do so, they had to develop tools and procedures for working with individual DNA molecules in order to make the transformations they desired. One of the first steps in

FIGURE 1.6
The Genetic Code

		2nd base			
		T	C	A	G
1st base	T	TTT = Phe TTC = Phe TTA = Leu TTG = Leu	TCT = Ser TCC = Ser TCA = Ser TCG = Ser	TAT = Tyr TAC = Tyr TAA = STOP TAG = STOP	TGT = Cys TGC = Cys TGA = STOP TGG = Trp
	C	CTT = Leu CTC = Leu CTA = Leu CTG = Leu	CCT = Pro CCC = Pro CCA = Pro CCG = Pro	CAT = His CAC = His CAA = Gln CAG = Gln	CGT = Arg CGC = Arg CGA = Arg CGG = Arg
	A	ATT = Ile ATC = Ile ATA = Ile ATG = Met/START	ACT = Thr ACC = Thr ACA = Thr ACG = Thr	AAT = Asn AAC = Asn AAA = Lys AAG = Lys	AGT = Ser AGC = Ser AGA = Arg AGG = Arg
	G	GTT = Val GTC = Val GTA = Val GTG = Val	GCT = Ala GCC = Ala GCA = Ala GCG = Ala	GAT = Asp GAC = Asp GAA = Glu GAG = Glu	GGT = Gly GGC = Gly GGA = Gly GGG = Gly

that direction arose from the work of the Swiss microbiologist Werner Arber and his colleagues. Arber was interested in the fact that bacteria appear to have evolved a mechanism for protecting themselves against attack by viruses (called *bacteriophages*, or just *phages*). That mechanism involves the use of enzymes with the ability to make scissor-like cuts in the DNA of the invading phage particles. Werner and his team were able to isolate the specific enzymes used by bacteria for this purpose, enzymes that were given the name *restriction endonucleases* or *restriction enzymes*.

This discovery, for which Arber received a share of the 1978 Nobel Prize for Physiology or Medicine, provided researchers with the first tool they needed in working with DNA molecules, a way of slicing open the molecule. The additional feature needed for that tool, however, was a "recognition" feature that could be used to cut a DNA molecule at any desired and specific point, such as the bond between an adenine nucleotide and a cytosine nucleotide ($A - C \rightarrow A + C$), or between a guanine nucleotide and thymine nucleotide ($G - T \rightarrow G + T$). That step was accomplished in 1970, when American molecular biologist Hamilton O. Smith and his colleagues discovered a restriction enzyme that recognizes a specific nitrogen base sequence in a DNA molecule. That enzyme, which they called *endonuclease R*, but is now known as *HindII*, "recognizes" the base sequence shown below and cuts it in the center of the sequence, as shown by the arrows.

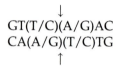

$$
\begin{array}{c}
\downarrow \\
GT(T/C)(A/G)AC \\
CA(A/G)(T/C)TG \\
\uparrow
\end{array}
$$

More than 3,000 restriction enzymes have now been discovered, each with the unique property of recognizing and cutting a DNA molecule within some specific base sequence. Many of these enzymes are commercially available for "off-the-shelf" use by researchers. With these enzymes, researchers now have a way of cutting apart a DNA molecule at virtually any point within its structure.

A "cutting tool," such as restriction enzymes, is essential in manipulating a DNA molecule since it provides a way of opening up the molecule. But another kind of tool is also necessary, one that seals the molecule once a desired change has been made in it. By the mid-1960s, scientists already knew that such tools must exist in nature. That knowledge comes from the fact that DNA molecules have the ability to repair themselves after being damaged. For example, DNA that has been damaged by exposure to X rays is often found later to have been repaired by some mechanism. That mechanism, researchers decided, must be some kind of enzyme that "knits up" the broken pieces of DNA and makes it functional again. The search was on for that enzyme.

In 1967, a molecular geneticist at the National Institutes of Health, Martin Gellert, reported that his research team had found the putative compound, an enzyme known as a ligase (from the Latin, *ligare*, "to bind"). More specifically, since this enzyme repairs damage to DNA molecules by restoring bonds that have been broken, the compound is known as DNA ligase. The discovery of ligases was of significance, because researchers now had the tools both to cut DNA molecules (restriction enzymes) and to put the molecules back together again (DNA ligases).

The first successful use of these tools in the manipulation of a DNA molecule was accomplished in 1972 by a research team headed by American biochemist Paul Berg. Berg worked with two well-studied viruses, the SV40 (for "simian virus 40") monkey virus and a bacterial virus known as the λ (lambda) bacteriophage. The DNA in both viruses consists of closed loops. The first step in this experiment, as shown in Figure 1.7, involved cutting open the phage DNA with a restriction enzyme, converting it into a linear molecule. A section of the phage DNA was then removed and treated at both ends with chemical groups that made it "sticky"; think of a small strip of Velcro at each end of the strip. Another restriction enzyme was used to cut open the SV40 viral loop, and Velcro-like groups were also added to the ends of the open loop. The "sticky" strip of phage DNA was then attached to opposite ends of the SV40 open chain, and the SV40-λ combination was sealed up by treatment with DNA ligase and a variety of other enzymes. The product of this experiment was a single molecule of DNA consisting of base sequences from two different organisms. It was a *recombinant*

FIGURE 1.7
Making Recombinant DNA

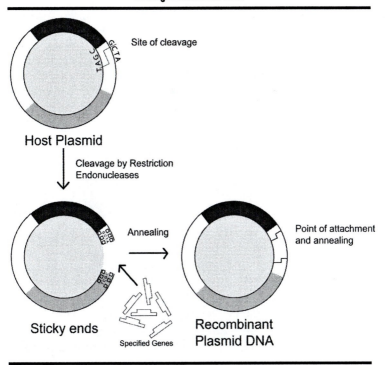

DNA (rDNA) molecule, the first ever made by humans. The molecule could also be described in other ways, as a chimera, for example, or as a transgenic molecule. The word *chimera* comes from the Greek mythological animal with the body and head of a lion, the tail of a snake, and a goat's head protruding from its spine. The term *transgenic*, similarly, refers to any organism or material resulting from one or more genes from a foreign organism being transplanted into a host organism.

The next step beyond Berg's research was already underway by the time his results were published. At Stanford University, medical researcher Stanley N. Cohen was studying bacteria that are resistant to antibiotics. He knew that bacteria must have this property because of the action of genes stored on their DNA. In the case of bacteria, this DNA exists in the form of circular loops known as *plasmids*. How genes on the plasmid confer antibiotic resistance to bacteria was his research problem. At virtually the same time, Herbert Boyer at the University of California at San Francisco was working on a line of research similar to that of Berg's, attempting to learn more about the function of restriction enzymes in the common bacterium *Escherichia coli*, usually called simply *E. coli*.

Cohen and Boyer did not know each other and probably had heard little or nothing about each other's work. That situation changed, however, when the two men were both invited to present papers about their work at a joint U.S.–Japan conference on bacterial plasmids held in Honolulu in November 1972 . Each was interested to hear about the other's research, and over sandwiches at a local delicatessen, they discussed the possibility of collaborating on a new project. After returning to California, they began a series of experiments that drew on the skills each had developed in working with microorganisms.

Their first experiment was relatively simple in concept, although a technological challenge. They worked with a plasmid that had been invented by Cohen, the pSC101 plasmid (*p* for "plasmid"; *SC* for "Stanley Cohen"; and *101*, an identifying number). The pSC101 is a very simple plasmid consisting of only two genes. One gene codes for replication of the plasmid, and the second gene codes for resistance to the antibiotic kanamycin. The value of the second gene is that it allows researchers to identify the presence of the gene in an experiment by treating the organisms present with the antibiotic. Using Boyer's expertise, they cut the plasmid with a restriction enzyme known as *Eco* RI and inserted a new gene into the opened plasmid, a gene that codes for resistance to the antibiotic tetracycline. They then used DNA ligase and other enzymes to seal up the plasmid, which now had three genes: one for replication, one for kanamycin resistance, and one for tetracycline resistance. The altered plasmid was then inserted into a colony of *E. coli* bacteria and kanamycin and tetracycline were added to the colony. Boyer and Cohen found that some bacteria were killed off by the kanamycin and some by the tetracycline. But some remained healthy, indicating that they had absorbed the altered plasmid into their own genetic material. The experiment demonstrated that it was possible to change the genetic characteristic of an organism (*E. coli* bacteria in this case) by adding a new component (the tetracycline gene in this case) to the host organism.

The next series of Boyer-Cohen experiments was even more impressive. Using the same procedure as described above, they added a gene removed from the South African toad (*Xenopus laevis*) to the pSC101 plasmid and, again, inserted the plasmid into a colony of *E. coli* bacteria. After a period of time, they found that the toad gene was expressing itself in the *E. coli* colony, and that it continued to do so generation after generation. Perhaps of equal importance was the consequence of this experiment. Bacteria reproduce quite rapidly; in the case of *E. coli* reproduction occurs about every 20 minutes. That means that any substance produced by the normal metabolism of the bacteria—the *X. laevis* gene product in this case—could be captured as a product of the bacterium's normal process of reproduction. The bacterium had become a "factory" for the substance coded for by the toad gene (in this case, nothing other than an RNA molecule).

The Boyer-Cohen experiments have become a classic landmark in the history of DNA technology. Cohen chose to remain at Stanford and continue his research on antibiotic resistance in bacteria. Boyer, however, chose a somewhat different future. In 1976, he and venture capitalist Robert A. Swanson founded Genentech, a company whose purpose was to conduct basic research on recombinant DNA products and to develop those products for commercial use. In that respect, Genentech was the first company founded to commercialize the products of recombinant DNA research. Before it was a year old, the company had developed its first commercial product, genetically engineered somatostatin, a naturally occurring growth hormone inhibiting hormone that inhibits the production of somatotropin, insulin, gastrin, and other hormones.

Boyer was not the only person to recognize the potential commercial value of his discovery with Cohen. At Stanford University, Neils Reimer, then director of the university's newly established Office of Technology Licensing, tried to convince Cohen to file a patent on the Boyer-Cohen invention, if such it could be called. At first, Cohen was reluctant to seek financial advantage from a scientific discovery. Eventually he relented, however, and three patent applications were filed in 1974, one for the methodology used in the research and two for any products that might result from use of the procedure (one for prokaryotes and one for eukaryotes). The applications were filed in the name of the university. The U.S. Patent Office agonized for a long time over the question as to whether living organisms and procedures related to them could be patented, but eventually decided to issue the patents, which it did on December 2, 1980; August 28, 1984; and April 26, 1988. The three patents all expired in 1997. By the end of 2001, Stanford and the University of California had realized more than $255 million in profits from licenses to 468 companies. A significant amount of that profit came from 2,442 products that had been developed from the patented procedures. Licensing fees were a modest cost to the companies who had to pay them, considering that the 2,442 products had generated an estimated $35 billion is sales over the life of the patents (Beera 2009, 760–761).

Cloning

An important consequence of the Boyer-Cohen experiments was their possible adaptation for cloning. Cloning is the process by which an exact copy is made of a gene, a cell, or an organism. Some of the earliest research on cloning dates to the studies of the German embryologist Hans Spemann in the 1920s; he developed fundamental methods for cloning cells and organisms. Later, researchers found ways to clone tadpoles (Robert Briggs and Thomas King; 1952), carrots (F. E. Steward; 1958), frogs (John Gurdon; 1962), and mice (Karl Illmensee; 1977). But Boyer and Cohen provided a method for cloning at the most fundamental level. When a specific segment of DNA is inserted into a bacterium and the bacterium begins to reproduce, it coincidentally makes exact copies of the original DNA segment. After many generations, the bacterium will have produced dozens, hundreds, or thousands of exact copies—clones—of the original DNA segment.

Over the past three decades, methods of cloning genes, cells, and whole organisms have improved significantly, resulting in the production of cloned sheep, mice, monkeys, cows, cats, deer, dogs, horses, mules, oxen, rabbits, and rats (National Human Genome Research Institute 2009). The technology has also been used in an attempt to help save endangered species and, remarkably, to produce clones of species that have already gone extinct, raising at least the theoretical possibility that those species might once more be returned to life on Earth (see, for example, Choi 2009; "Cloning: Bringing Back Endangered Species" 2000).

Transgenic Plants and Animals

The work of Berg, Boyer, Cohen, and their colleagues demonstrated the feasibility of creating transgenic organisms, organisms with DNA from two or more different organisms. During the late 1970s and 1980s, researchers carried out a number of experiments designed to confirm this prediction and to create ever more complex transgenic bacteria, plants, and animals. An example of this research was the work of the German-American biochemist Rudolf Jaenisch. In the early 1970s, Jaenisch discovered a method for inserting genes in the early embryo of a mouse. When that embryo developed and became an adult, its own DNA had incorporated the foreign genes injected during the embryonic state. A 1974 paper by Jaenisch reported on the production of this research, the production of the first genetically engineered animal (Jaenisch and Mintz 1974, 1250–1254). Over the next three decades, researchers produced an almost bewildering variety of transgenic animals, including transgenic pigs, sheep, dairy cattle, chickens, fish (such as the GloFish®), and laboratory animals, such as mice and rats (for a review of these developments, see Melo, et al. 2007, 47–61).

Research on transgenic plants was also moving forward at the same time. In a somewhat unusual event, the first transgenic plants were introduced by four different research groups within a few months of each other. Three of those groups presented papers about their discoveries at a conference in Miami, Florida, in January 1983, while the fourth group announced its own discovery at a conference in Los Angeles in April of the same year. The three groups that reported in January all used similar approaches for inserting a gene providing resistance to the antibiotic kanamycin in plants. Two of the teams used tobacco plants, and the third, petunia plants (Fraley, Rogers, and Horsch 1983, 211–221; Framond, A. J. 1983, 159–170; Schell, et al. 1983, 191–209). The fourth group took a somewhat different approach and introduced a gene removed from the common bean plant and inserted it into a sunflower plant (Murai, et al. 1983, 476–482).

One of the interesting questions raised by research on transgenic plants and animals was whether life forms could be patented. When Stanford submitted applications for the procedures and products developed by Boyer and Cohen in 1974, the U.S. Patent Office (now the U.S. Patent and Trademarks Office; USPTO) was uncertain as to how it should respond. It had never before been asked to patent living organisms, the products of their metabolism, or the procedures by which they functioned. As noted above, it waited six years before making its first decision about this question. One issue involved in its decision was how the courts would act on applications of this kind. That roadblock was removed in 1980 when the U.S. Supreme Court ruled that the Patent Office could issue a patent to the General Electric Company for an organism invented by the Indian-American chemist Ananda Chakrabarty. Chakrabarty had used recombinant DNA technology to develop a bacterium from the *Pseudomonas* genus that was capable of breaking down crude oil. He anticipated that the bacterium would be useful in cleaning up oil spills. At first, the Patent Office rejected Chakrabarty's application for a patent because it did not think that living organisms could be patented. Chakrabarty and General Electric appealed the Patent Office's decision until the case reached the Supreme Court. On June 16, 1980, Chief Justice Warren Burger wrote that "A live, human-made micro-organism is patentable subject matter under [section] 101 [of Title 35 of the U.S. Code]. Respondent's micro-organism constitutes a 'manufacture' or 'composition of matter' within that statute" (*Diamond v. Chakrabarty* 1980, 446). In reaching this decision, Burger (and the court majority) reviewed the position of the U.S. Congress on the issuance of patents for plants in its discussions prior to adopting the Plant Patent Act of 1930. At that time, the position of the Congress was that

> There is a clear and logical distinction between the discovery of a new variety of plant and of certain inanimate things, such, for example, as a new and useful natural mineral. The mineral is created wholly by nature unassisted by man . . . . On the other hand, a plant discovery resulting from cultivation is unique, isolated, and is not repeated by nature, nor can it be reproduced by nature unaided by man . . . . (Senate Report No. 315 and House Report No. 1129, as cited in *Diamond v. Chakrabarty* 1980, 447)

From these statements, Burger concluded that

> Congress thus recognized that the relevant distinction was not between living and inanimate things, but between products of nature, whether living or not, and human-made inventions. (*Diamond v. Chakrabarty* 1980, 447)

The court's decision was of profound significance, because it meant that a researcher could apply for a patent on virtually any novel living organism, the method by which it was produced, the tools and techniques developed for its production, and any products that might be obtained from the organism. The first beneficiaries of the court's 1980 decision were researchers working with bacteria, viruses, and other lower forms of life. Before long, however, researchers working with organisms at every level—all kinds of plants and animals—had begun to seek patents for their inventions and discoveries. The DeKalb Genetics company, for example, sought a patent for a genetically engineered corn plant with a greater than average level of the amino acid tryptophan, improving its nutritional value for animals to whom it was fed. The USPTO rejected DeKalb's request, a decision that was later overruled by the Board of Patent Appeals. The patent was issued in 1990, the first patent for a transgenic plant.

At about the same time, researchers at Harvard University were developing the oncomouse, a mouse that has been engineered to carry an additional gene that places the animal at significantly greater risk for cancer than its non-engineered cousins. The value of the mouse is that it can be used in cancer research. The USPTO issued a patent for the oncomouse on April 12, 1988, and its name was registered as the OncoMouse® by the DuPont company, which holds licensing rights to the invention. The inventors also applied for a patent in Europe in June 1985, and after a very long series of denials and appeals, the patent was granted on July 6, 2004. Only in Canada has the application process failed. On December 5, 2002, the Canadian Supreme Court ruled that the mouse was not an "invention" and that it could not, therefore, be patented in Canada ("Bioethics and Patent Law: The Case of the Oncomouse" 2006).

The legality of patenting life forms continued, understandably, to be a matter of dispute. From time to time, cases arose that questioned the right of a company to patent a plant, an animal, a protist, a gene, or some other product or methodology associated with recombinant DNA manipulation of living organisms. In 2001, the U.S. Supreme Court again considered this issue. J.E.M. Ag Supply, Inc., doing business as Farm Advantage, Inc., had sued Pioneer Hi-Bred International, Inc., a company that held 17 patents on engineered plants. J.E.M. Ag Supply claimed, the court's 1980 decision notwithstanding, that genetically engineered plants could not be patented. The case reached the court on October 3, 2001, and was settled on December 10, 2001. The court, in a six to two ruling, reaffirmed its position, stating that nothing in existing federal law prohibited the patenting of novel organisms, their parts, or technology used to produce them (*J.E.M. Ag Supply, Inc., DBA Farm Advantage, Inc., et al. v. Pioneer Hi-bred International, Inc. 2001*).

With the legal status of patented life largely settled in the United States, the USPTO has been deluged with applications for patents on almost every conceivable type of recombinant product. Today it is difficult to estimate the number of patents issued by the USPTO in the last three decades in this area. A search of the USPTO database for terms related to such patents granted since 1976 produces very large numbers of patents, more than 73,000 for "organisms," more than 113,000 for "genes," more than 193,000 for "animals," and more than 210,000 for "plants." It appears that almost every development in the engineering of life forms has become and will become the property of some individual or corporation.

Practical Applications of Transgenic Organisms

The work of Berg, Boyer, Cohen, and countless numbers of researchers in the late 1950s, 1960s, and early 1970s provided a bridge between the "what we think can be done" occasioned by the discovery of the DNA structure by Watson, Crick, and Franklin, and the "what CAN be done" of today's DNA technology. The products of today's DNA technology fall primarily into four categories: research, pharmaceuticals, agriculture, and industry.

Research

The OncoMouse® developed by Harvard researchers Philip Leder and Timothy A. Stewart in the late 1980s is an early example of the way a genetically modified organism can be used in research. Since the OncoMouse® is engineered to contain an oncogene, a gene that tends to cause cancer in an organism, the engineered organism provides an ideal system for the study of cancer and methods that can be used for its treatment. A somewhat different example is the knockout mouse, first created by Italian-American molecular geneticist Mario R. Capecchi, British developmental biologist Martin Evans, and British-American geneticist Oliver Smithies in 1989 (for which they won the 2007 Nobel Prize in Physiology or Medicine). The knockout mouse is a mouse that has been genetically modified by the inactivation of one or more genes. This change allows a researcher to determine precisely the function of some particular gene that has been "knocked out."

Pharmaceuticals

One of the most productive uses of recombinant DNA technology has been the use of bacteria for the production of pharmaceuticals. The procedure used in this field is identical to that employed by Berg, Boyer, and Cohen. First, a gene coding for the production of some desired protein is obtained, either by removing it from an organism or by synthesizing it in the laboratory. That gene is then inserted into a plasmid (as in the Berg and Boyer-Cohen experiments), a virus, or some other structure, called a *vector*. The vector is then inserted into a host cell, most commonly a bacterium. If all goes well, the bacterium incorporates the engineered plasmid into its genetic material and begins to produce the protein coded by the inserted gene along with its normal metabolic products. When the bacterium reproduces, it produces two copies of itself (then four, then eight, then 16, and so on), each of which continues to produce the desired protein. A large vat of bacteria becomes, then, a microbial factory for the generation of the desired protein product.

This method has now been used to produce hundreds of different therapeutic proteins, more than 130 of which have been approved for commercial use by the U.S. Food and Drug Administration (FDA). Table 1.1 summarizes some of the drugs that have been prepared by rDNA technology and have been approved for use by the FDA (for a review of the current status

TABLE 1.1
Some Genetically Engineered Drugs Approved by the FDA

Drug	Year Approved[1]	Approved Use
Human insulin	1982	Diabetes mellitus
Human growth hormone	1985	Growth deficiency in children
Alpha interferon (2b)	1986	Hairy cell leukemia; genital warts; Kaposi's sarcoma; hepatitis B; hepatitis C
Hepatitis B vaccine	1986	Hepatitis B
Alterplase	1987	Acute myocardial infarction; acute massive pulmonary embolism
Epoetin alfa (erythropoietin)	1989	Anemia associated with chronic renal failure
Gamma interferon (1b)	1990	Chronic granulomatous disease; severe, malignant osteopetrosis.
Antihemophiliac factor	1992	Hemophilia A
Proleukin (IL-2)	1992	Metastatic kidney cancer; metastatic melanoma
Beta interferon (1b)	1993	Certain types of multiple sclerosis
Imiglucerase	1994	Gaucher's disease
Pegaspargase	1994	Acute lymphoblastic leukemia
Coagulation factor IX	1997	Factor IX deficiencies
Glucagon	1998	Hypoglycemia in diabetics
Adalimumab	2003	Rheumatoid arthritis
Platelet-derived growth factor-BB	2005	Periodontal bone defects and associated gingival recession
Hyaluronidase	2005	As an adjunctive to other injected drugs
HPV vaccine	2006	Human papillomavirus disease
Sorafenib	2007	Primary kidney cancer; primary liver cancer

[1]Additional approvals may be given for similar or related products at a later date.

of the manufacture of drugs by genetically modified organisms, see Leader, Baca, and Golan 2008, 21–39).

Bacteria provide a simple and inexpensive mechanism for the production of recombinant pharmaceuticals. But they are by no means the only way of making drugs with the use of rDNA technology. For nearly two decades, researchers have been considering the use of farm animals for the production of useful proteins. Animals have a number of advantages over bacteria for the synthesis of human proteins. They have a built-in system for making proteins that operates at the correct temperature and pH (acidity), and a natural immune system provides an ideal environment for the culturing and harvesting of drugs produced from transplanted genes. They also have all the nutrients needed to keep the system operating efficiently, and they possess a natural output system in the form of urine, blood, and/or milk from which manufactured drugs can be collected.

One of the pioneers in the field was PPL Therapeutics, the company that was responsible in 1996 for the development of the first cloned mammal, the sheep known as Dolly. In 1990, PPL transplanted a gene for the production of the protein apha 1 antitrypsin (AAT) into the DNA of a transgenic sheep by the name of Tracy. AAT is normally produced in the liver and transported to the lungs, where it is used to support breathing. Individuals who lack the protein are short of breath and have trouble breathing when carrying out even the simplest activities.

The experiment appeared to be very successful. More than half of the protein in milk produced by Tracy was AAT. And her descendants were equally productive. By 1998, PPL reported that Tracy had more than 800 granddaughters who were producing an average of about 15 grams per liter of AAT per day. That success was short-lived, however, as clinical trials with humans using the engineered protein found that a number of patients developed an unexplained, but troublesome, "wheezing" problem. PPL decided to discontinue clinical trials of the engineered AAT, and the trials have not since been restarted (Colman 1999, 167–173).

The problems experienced with Tracy have not discouraged other researchers or pharmaceutical corporations from pursuing the use of farm animals for the production of drugs, a field now commonly known as *pharming*, for "pharmaceutical development using farm animals." This technology is also widely described as *plant molecular farming* (PMF), or simply, *molecular*

TABLE 1.2
Some Pharmed Products Currently under Development

Pharmaceutical	Host Animal	Target Disorder(s)
Alpha-1-antitrypsin	Sheep	Emphysema; cystic fibrosis; other lung disorders
Cystic fibrosis transmembrane conductance regulator (CFTR)	Sheep	Cystic fibroses
Tissue plasminogen activator	Sheep, pig	Thrombosis
Factor VIII and factor IX	Sheep, pig, cows	Hemophilia
Fibrinogen	Sheep, cows	Wound healing
Human protein C	Goat	Thrombosis
Collagen I and collagen II	Cow	Anti-arthritic; tissue repair
Human serum albumin	Cow, goat	Surgery, burns, shock
Human fertility hormones	Cow, goat	Human infertility
Lactoferrin	Cow, maize	Gastrointestinal tract disorders; anti-arthritic
Erythropoietin	Rabbit	Anemia associated with dialysis
Interleukin-2	Rabbit	Renal cell carcinoma
Human growth hormone	Rat	Pituitary dwarfism
Gastric lipase	Maize	Cystic fibrosis; pancreatitis
Hepatitis B vaccine	Lettuce, potato	Hepatitis B
Norwalk virus vaccine	Potato	Norwalk virus disease
Insulin	Safflower	Diabetes
Cytokine (transforming growth factor-beta 2)	Tobacco	Ovarian cancer

Source: Some of the information in this table comes from *Pharmaceutical Biotechnology*, by Daan J. A. Crommelin, Robert D. Sindelar, Bernd Meibohm, 154.

farming. In addition, researchers have expanded potential host organisms to include plants. For example, researchers at the Agragen corporation are studying the possibility of using genetically altered flax plants to manufacture human serum albumin and omega-3 fatty acids (Boyer 2009). Some of the pharmed drugs currently under development are listed in Table 1.2.

An important breakthrough occurred in 2006 when the FDA gave approval for the use of the first product from a genetically altered mammal. The product is ATryn®, a recombinant form of human antithrombin III, a small protein that prevents excessive clotting of blood. The product is made by the GTC Biotherapeutics company ("ATryn—Recombinant Human Antithrombin" 2009).

Agriculture

As noted above, humans have for centuries used selective breeding to improve the quality of the crops and animals they grow.

They have tried to develop cows that produce more milk, sheep with better wool, stronger horses, corn that will be more resistant to disease, and larger, more nutritious fish. DNA technology has made it possible for plant and dairy farmers to achieve these same objectives more rationally and more efficiently. For example, PPL Therapeutics announced in February 1997 that it had developed a transgenic cow named Rosie whose milk contained the human protein alpha-lactalbumin. The protein was coded for by a gene inserted into Rosie's own DNA when she was still at the embryonic stage. The presence of the human protein in Rosie's milk made it nutritionally superior for humans than is natural cow's milk. PPL hoped to market the transgenic milk as a substitute for infant formula or as a nutritional supplement for adults with special dietary or nutritional needs (Begley 1997).

Some of the characteristics for which transgenic animals are produced are the following:

- Increased rate of growth;
- Greater muscle mass and/or bone strength;
- Greater resistance to pain;
- Leaner, more tender pork and beef;
- Resistance to disease;
- Increased production of milk and/or for longer periods of time;
- Improved nutrition of eggs (for chickens);
- Milk that lacks human allergens;
- Milk that produces increased amounts of cheese and yogurt;
- Modifications in physical conditions that increase the efficiency of farming, such as the production of hornless cattle;
- Improved feeding of stock, as in increasing the ability of animals to digest cellulose;
- Fish that grow faster than conventional fish;
- Fish that reach sexual maturity at a younger age than conventional fish;
- Fish that tolerate a wider variety of environmental conditions, such as colder or warmer water.

One example of an apparent success in this line of research was the 2005 announcement of the development of a mastitis-resistant cow developed by researchers at the U.S. Department of Agriculture. Mastitis is an infection of the udder, a common and potentially serious disease among dairy cows, significantly affecting their milk production. Vaccines and antibiotics are currently available for the prevention and treatment of mastitis, but they are not very effective. In the USDA experiment, researchers constructed a gene that codes for the protein lysostaphin, which kills the bacterium that causes mastitis. They then inserted the gene into the DNA of a number of cow embryos. When those cows grew to maturity, their milk contained the lysostaphin protein, and they had a dramatically lower rate of mastitis infections than did a group of control cows.

The development of transgenic animals for agricultural applications is still at the development stage. One reason that progress has been slow is that the addition of a foreign gene to an animal's natural genome sometimes has unexpected and serious side effects. Perhaps the most frequently cited case of such side effects is the so-called Beltsville pigs from the late 1980s. A group of pig embryos received human growth hormone in an attempt to increase their rate of growth and reduce body fat. The experiment appeared to be at least partially successful as the mature pigs' body fat was reduced and their feeding efficiency did improve. However, 17 of the 19 pigs who had received the gene died in the first year, and all animals exhibited a number of debilitating disorders, including diarrhea, mammary development in males, lethargy, arthritis, lameness, skin and eye problems, loss of libido, and disruption of estrous cycles. Because of these problems, the experiment was discontinued (Committee on Defining Science-Based Concerns Associated with Products of Animal Biotechnology. Committee on Agricultural Biotechnology, Health, and the Environment. National Research Council 2002, 98).

A somewhat different use of transgenic animals is in xenotransplantation, the transplantation of an organ or other body part from one species to a different species. Over the past 30 years, a number of experiments have been conducted involving the transplantation of hearts, lungs, kidneys, livers, and cells from mammals, such as rabbits, pigs, goats, sheep, and various nonhuman primates. The most common problem with this type of procedure is that the human host body usually rejects the transplanted organ. The body's immune system recognizes the pig, sheep, cow, baboon, or other organ as a foreign object and mounts its full immune arsenal to destroy the invader, almost always with success. One possible solution to this problem is to engineer the donor animal's DNA so that it more closely resembles that of a human. In an example of this type, Jeffrey Platt at Duke University has explored the possibility of using genetically engineered pig livers as "bridge" (temporary replacement) organs for patients with life-threatening renal failures waiting for a human transplant (Frederickson 1997, 439). Platt reported in 2000 that two patients who had received pig livers transformed to include human genes designed to reduce rejection had survived for periods of five and 18 months (Levy, et al. 2002, 272–280).

DNA technology has also made it possible to modify plants, so as to introduce a number of desirable properties: resistance to disease, protection against attack by insects, improved nutritional characteristics, and reduced rates of spoilage, for example. One of the first genetically modified (GM) agricultural products to be developed was a biological control agent called Frostban. The agent was developed as a way of protecting crops from freezing. Scientists have long known that crops will not freeze even when the temperature drops below 0°F in the absence of certain frost-promoting bacteria. Frostban was a genetically altered form of the common Pseudomonas bacterium that attacked and destroyed those bacteria. Developed and field tested in the early 1980s, Frostban never reached the marketplace because of some serious problems with late field tests and approval issues from the federal government.

Another early example of failed GM products was the Flavr Savr™ tomato developed by Calgene, Inc., in the early 1990s. The tomato contained a gene that interferes with the production of the enzyme polygalacturonase, which causes a tomato to soften. Lacking this enzyme, a tomato could remain on the vine and ripen before being picked, in contrast to conventional tomatoes, which are usually picked while they are still green. The Flavr Savr™ tomato performed well in field tests, and Calgene requested approval from the FDA in 1991 for its commercial sale. In May 1994, the FDA granted its approval, noting that the Flavr Savr™ was "as safe as tomatoes bred by conventional means" (U.S. Food and Drug Administration 1994). Calgene began marketing the Flavr Savr™ tomato in 1994 but had little success with the product. Critics have pointed to a number of reasons for its failure in the marketplace, including the reluctance of consumers to purchase genetically modified foods and a poorly conceived marketing plan. In any case, it was withdrawn from the market shortly after Calgene was purchased by Monsanto in 1995 (Soil Association 2005).

Since the 1990s, researchers have experienced far more success in the development of genetically engineered crops with a range of desirable properties. These products tend to fall into three categories outlined below.

GM Products Resistant to Certain Pests

Some examples of these products include GM corn that is resistant to the European corn borer, stalk borer, southwestern corn borer, armyworm, and corn earworm; GM cotton that is resistant to the cotton bollworm, the pink bollworm, and the tobacco budworm; and GM potato that is resistant to the Colorado potato beetle. These products are sometimes scientifically successful, but commercially a failure. For example, the GM potato described here, marketed under the brand name of New-Leaf, was approved for sale by the FDA, but was withdrawn from the marketplace in 2001 because of poor sales.

GM Products Resistant to Herbicides

Herbicides are widely used by farmers to control weeds that compete with their crops. One of the best known and most widely used herbicides is Roundup®, manufactured by Monsanto. The most serious problem with such herbicides is that they often kill crops along with the weeds. The solution to this problem has been to develop, by DNA technology, crops with foreign genes that protect them against herbicides. One of the most successful of these new products is so-called "Roundup Ready®" crops, crops that are not harmed when sprayed with Roundup®. Today, seeds for a number of Roundup Ready® crops are available from Monsanto, including those for cotton, corn, canola, lettuce, sugar beets, soybeans, tobacco, tomato, and wheat. GM crops resistant to herbicides other than Roundup® are also under development or in use for these and other crops, including alfalfa, barley, flax, melons, peanuts, potatoes, rice, sweetgum, tobacco, and tomato ("Herbicides and Herbicide Resistance in Transgenic Plants" 2009).

GM Products with Nutritional Value or Other Benefits

In the 1990s, Swiss botanist Ingo Potrykus began research on a genetically modified form of rice capable of synthesizing beta carotene, a precursor of vitamin A. Potrykus was interested in this problem because vitamin A deficiency is responsible for an estimated 250,000–500,000 cases of blindness in children around the world. If he could improve the nutritional quality of the primary food for the vast majority of those children, he reasoned, he could significantly reduce this serious public health problem. The product he invented, called *golden rice*, is a form of conventional white rice that includes three inserted genes that code for beta carotene. It is the first genetically engineered product made specifically to improve the nutritional value of a natural food. The method Potrykus used, however, is applicable in principle to a comparable treatment of virtually any natural food.

One of the increasing serious problems faced by farmers around the world is insufficient land area on which to grow their crops. More and more, they are using land that lacks sufficient water, has too high a salt content, or is unsatisfactory for some other reason. For over a decade, researchers have been looking for ways to genetically modified crops to be more tolerant of drought and high salt concentrations. One approach has been to insert the gene for the enzyme H^+-pyrophosphatase, which appears to make plants more resistant to drought and to sodium chloride (NaCl). In one series of experiments along this line, engineered plants with the additional genes for H^+-pyrophosphatase were able to survival longer with water and better when exposed to high concentrations of salt than were conventional plants (Gaxiola, et al. 2001, 11444–11449). Research on drought- and salt-tolerant maize and oilseed rape is already well advanced, and Monsanto has announced that its comparable corn product will be ready for the market in 2012 ("Frontrunners in Securing and Increasing Yield in Crops" 2009).

Industry

In 2002, researchers from Nexia Biotechnologies Inc. and the U.S. Army Soldier Biological Chemical Command (ASBCC) announced that they had manufactured silk fibers normally produced by spiders using recombinant DNA technologies. That discovery was important because, while spider silk is generally recognized as the strongest fiber in the world, no commercial system for extracting the material from spiders has ever been developed. The Nexia and ASBCC researchers had achieved this goal by inserting spider genes controlling the synthesis of silk into goats. Milk produced by the goats contains the spider silk protein, which can then be separated, purified, and spun into threads. The first applications of this GM material, called Bio-Steel®, will probably be used in the manufacture of bulletproof vests and anti-ballistic armor for the U.S. military.

Industrial applications of DNA technology have not developed as rapidly as have those in pharmaceuticals and agriculture, but they hold as much potential. Just a few of the myriad possible products that may result from this line of research are:

- Modifying fats and oils produced by plants to obtain products for use in paints and manufacturing processes;
- Production of polymeric materials from corn plants for the production of plastics;
- Modifying the genetic structure of plants to make them sensitive to certain external stimuli, making them useful as biosensors;
- Engineering of plants to make possible the synthesis of artificial forms of natural and synthetic fibers, such as cotton, wool, and nylon;
- Modifying flowering plants to give them more intense and/or more interesting flower colors and odors;
- Production of enzymes that can be used to convert biomass into fuels more efficiently than current methods.

Gene Therapy and Genetic Testing

Two other fields of medicine that have benefited significantly from the development of DNA technology have been gene therapy and genetic testing. Gene therapy (also known as human gene therapy) is a procedure by which genes are inserted into the body of a person with a genetic disorder. Researchers first began to appreciate the potential usefulness of gene therapy in the 1970s when recombinant DNA procedures were first being developed. Gene therapy involves five basic steps:

1. The gene responsible for a genetic disorder must be identified. A genetic disorder develops because a patient lacks an essential gene or has a faulty copy of that gene, such that some essential protein is either not made or is made incorrectly. For example, the disorder known as

phenylketonuria (PKA) occurs because a person lacks the gene needed to make the enzyme hepatic phenylalanine hydroxylase. Without this enzyme, the body cannot metabolize the amino acid phenylketonuria, and a disease develops.

2. A correct copy of that gene must be obtained from some natural source or must be produced synthetically in the laboratory.

3. That gene must be introduced into the patient's DNA. rDNA procedures for splicing a gene into a DNA molecule developed in the 1970s provided a mechanism for inserting a correct gene into a patient's DNA.

4. The altered DNA must then be introduced into cells from the patient's body. Originally, this step was performed by removing a sample of blood from the patient's body and inserting corrected DNA into lymphocytes (white blood cells) in the blood. A virus was typically used as a vector (carrier) for the inserted gene, although alternative vectors and methods of insertion are now available.

5. Cells containing the altered (correct) DNA are then returned to the patient's body where, if the procedure is successful, they begin to synthesize the missing protein. They must also continue to do so as the lymphocytes (or other engineered cells) reproduce over and over again in the patient's body.

After more than five years of preparation, the first gene therapy trial was conducted on September 14, 1990, by a group of researchers led by American medical researcher and molecular biologist W. French Anderson. Their subject was a four-year-old girl with adenosine deaminase (ADA) deficiency, a condition that leaves a person susceptible to all kinds of infections. The condition is usually fatal at an early age. Anderson's team followed the general procedure described above, providing the patient with the gene needed to synthesize the adenosine deaminase enzyme she was lacking. Although the procedure had to be repeated a number of times following the original experiment, it appeared to have been at least partially successful, and the patient remains alive today.

Gene therapy has been plagued by a number of problems and tragedies since the earliest successes of Anderson's research. For example, 18-year-old Jesse Gelsinger died in 1999 while being treated for a rare liver disease called ornithine transcarbamylase disorder. In 2007, 36-year-old Julie Mohr died while being treated for rheumatoid arthritis, although the cause of her death may very well not have been the gene therapy treatment itself. Still, many researchers have great confidence that gene therapy will save the lives of countless humans once the technology used in the procedure is well developed. As of 2008, the latest year for which data are available, more than 1,340 gene therapy clinical trials worldwide have been conducted. The number of trials increased rapidly from 1990 to 1999, but then fell off significantly after the death of Jesse Gelsinger. Soon thereafter, they began to increase again, and have numbered about 100 worldwide since the early 2000s, with 95 in 2004, 98 in 2005, and 97 in 2006 (Edelstein, Mohammad, and Wixon 2007, 833–842; Erfo Centrum 2009).

Genetic Testing

Scientists have known for well over a century that certain diseases have a genetic basis. The appearance of hemophilia among the children of Queen Victoria in the mid-nineteenth century was a subject of considerable interest to medical researchers at the time. By the middle of the twentieth century, researchers had begun to find an association between specific types of genetic material, such as chromosomes, and specific genetic disorders. In 1959, for example, French geneticist Jerome Lejeune discovered that Down syndrome is caused by the presence of an extra copy of chromosome 21. Still, it took many years for scientists to take advantage of the Watson and Crick discovery and pinpoint specific alterations in DNA molecules that result in genetic disorders.

Efforts to develop methods for testing for genetic disorders date to 1961 when American microbiologist and physician Robert Guthrie developed a test for phenylketonuria (PKU), a genetic disorder in which the body is unable to metabolize the amino acid phenylalanine. The amino acid builds up in the body and may produce serious health effects and, in some cases, death. Once detected, however, the disease is easily treated by providing the patient with a diet low in foods that produce phenylalanine. Guthrie's test was very simple. A drop of blood is drawn from the heel of a newborn child and then tested for the presence of phenylalanine. The amount of the amino acid present in the sample tells whether or not the child has the disorder.

Four kinds of genetic disorders occur. Single-gene disorders are the most common. They occur when a change occurs in a single gene, or the gene is lost entirely. The loss of or damage to the gene means that it is unable to produce some protein essential to a person's well-being. Currently, more than 6,000 single-gene disorders are known. Multifactorial genetic disorders occur when more than one gene is damaged or lost and/or when an environmental factor is present that moderates the genetic error. Breast cancer, heart disease, diabetes, and hypertension are examples of multifactorial disorders. Chromosomal genetic disorders occur as the result of some change in a whole chromosome, as is the case with Down syndrome. Mitochrondrial genetic disorders are rare conditions that result from errors in the noncoding DNA found in mitochondria. Clearly, the easiest genetic disorders to test for are single-gene disorders, and the vast majority of genetic tests currently available fall into that category.

Until the late 1990s, only gross methods were available for genetic tests. Testing for Down syndrome, for example, involves taking a drop of blood and examining it under a microscope. The condition of chromosomes in the blood can easily be seen, and the presence of trisomy (three copies) on chromosome 21 can be detected. This method works well for any chromosomal genetic disorder. Detecting single-gene disorders is more difficult, however, since one cannot see the specific base sequence present on a chromosome microscopically. The detection of single-gene disorders was made much easier, however, when the complete sequencing of the human genome was completed in the early 2000s.

The Human Genome Project was a program initiated by the U.S. government in October 1990 to map the complete human genome, that is, to determine the exact base sequence for every gene on every human chromosome. That project was completed in 2003, at which point the base sequence for all 20,000–25,000 human genes was known. The value of this research is that scientists now know what the correct base sequence is for every gene in the human genome. Genetic tests can now be devised to determine the base sequence for any given gene from any given person and compare that result with the "normal" base sequence for that gene. If errors are present in the subject's base sequence, it becomes possible, at least in theory, to develop a method to correct that base sequence and to cure that disease. This principle forms the basis of much ongoing research in the development of new genetic tests.

Forensic DNA Testing

The applications of DNA technology thus far can all be characterized as developments in genetic research that have their origins from Mendel's original research in the mid-nineteenth century. Another application, however, was quite unexpected: the use of DNA testing in forensic science. The first breakthrough in this field came in 1984 when British geneticist Alec Jeffreys developed the restriction fragment length polymorphism (RFLP) test for DNA. That test makes use of the fact that a DNA molecule consists of two kinds of base sequences: exons and introns. An exon is a sequence of bases that codes for a protein. Exons tend to be essentially the same from one human (or other animal) to another because all humans need to make the same proteins to function normally. Interspersed between exons are other base sequences that do not code for a protein. In fact, they have no known function and are sometimes known for that reason as "junk DNA." Jeffreys found that introns differ from person to person. Although two individuals are likely to have identical exons, they are likely to have very different introns. He found a method for removing introns from an individual's DNA (using restriction enzymes) and labeling them with radioactive isotopes to permit a radiographic photograph of the removed segments.

Less than a year after this discovery, Jeffreys used the new technology to verify the paternity of a Ghanian boy who was applying for a passport to visit Great Britain. He compared the DNA "fingerprint" of the boy with that of a man claiming to be his father and found a match, making the boy eligible for entry into the country. A year later, Jeffreys was asked to use his technique in the investigation of a pair of rape-murders in the small town of Narborough in Leicestershire. Jeffreys was able to show that the man being held for the crimes was *not* guilty, but that some other unknown individual *was* responsible for both crimes. After further investigation by the police, a new suspect, Colin Pitchfork, was arrested, his DNA was tested by Jeffreys, and a match was obtained. Pitchfork thus became the first person convicted of a crime based on DNA testing ("Case Study #2: Colin Pitchfork" 2009).

Today, Jeffrey's system of DNA analysis by means of RFLP has been largely replaced by a second method of finding, cutting out, and identifying fragments of a DNA molecule. That method was invented by American molecular biologist Kary Mullis in 1986. The technology, called polymerase chain reaction (PCR) is preferable to RFLP, because it takes less time and can be used with much smaller samples of DNA. PCR gets its name from the fact that a very small sample of DNA—in principle, a segment of a single molecule of DNA—can be amplified many times over to produce millions of copies of the original sample, providing a much larger specimen with which to work.

The use of RFLP and PCR in forensic science is based on a statistical principle: DNA taken from two different individuals has only a small likelihood of being exactly the same. Mathematicians have developed models to predict what that likelihood is for various circumstances. In fact, whole books have been written on the subject. As a result, researchers are now able to look at samples taken from a crime scene and from a suspect and say that the chances are one in a million, one in a billion, one in ten billion, or some other number, allowing judges and juries to decide whether a particular individual is likely to have committed a particular crime. Forensic DNA analysis (sometimes called *DNA fingerprinting*) is now generally acknowledged as being the single most reliable tool for identifying criminal suspects throughout the world.

References

"ATryn®—Recombinant Human Antithrombin." URL: http://www.gtc-bio.com/products/atryn.html. Accessed on April 7, 2009.

Beera, Rajendra K. 2009. "The Story of the Cohen–Boyer Patents." *Current Science.* 96 (6, March 25): 760–761.

Begley, Sharon. 1997. "Little Lamb, Who Made Thee?" *Newsweek.* March 10, 1997. Available online. URL: http://www.newsweek.com/id/95479. Accessed on April 7, 2009.

"Bioethics and Patent Law: The Case of the Oncomouse." 2006. *WIPO Magazine.* 3 (June): 16–17. Available online. URL: http://www.wipo.int/wipo_magazine/en/2006/03/article_0006.html. Accessed on April 6, 2009.

Boyer, Mike. 2009. "Agragen Envisions Homegrown Medicine: Biotech Venture Aims to Produce Drugs in Crops." *Cincinnati Enquirer.* Available online. URL: http://www.enquirer.com/editions/2004/07/20/biz_biz2 agragen.html. Accessed on April 7, 2009.

"Case Study #2: Colin Pitchfork." 2009. Available online. URL: http://gohs.tvusd.k12.ca.us/TeacherWebs/Science/SFajardo/Web%20pages/Colin%20Pitchfork.htm. Accessed on April 8, 2009.

Choi, Charles Q. 2009. "First Extinct-Animal Clone Created." *National Geographic News.* February 10, 2009. Available online. URL: http://news.nationalgeographic.com/news/2009/02/090210-bucardo-clone.html. Accessed on April 8, 2009.

"Cloning: Bringing Back Endangered Species." *Applied Genetics News.* October 2000. Available online. URL: http://findarticles.com/p/articles/mi_m0DED/is_3_21/ai_66520544/. Accessed on April 8, 2009.

Colman, Alan. 1999. "Dolly, Polly and Other 'Ollys': Likely Impact of Cloning Technology on Biomedical Uses of Livestock." *Genetic Analysis: Biomolecular Engineering.* 15 (3–5, November): 167–173.

Committee on Defining Science-Based Concerns Associated with Products of Animal Biotechnology. Committee on Agricultural Biotechnology, Health, and the Environment. National Research Council. 2002. *Animal Biotechnology: Science Based Concerns.* Washington, D.C.: National Academies Press.

Crick, F. H. C. 1958. "On Protein Synthesis." *Symposia of the Society for Experimental Biology: The Biological Replication of Macromolecules.* 12 (1958): 152–153.

Diamond v. Chakrabarty, 447 U.S. 303 (1980). Available online. URL: http://supreme.justia.com/us/447/303/case.html. Accessed on April 6, 2009.

Edelstein, Michael L., Abedi R. Mohammad, and Jo Wixon. 2007. "Gene Therapy Clinical Trials Worldwide to 2007—An Update." *The Journal of Gene Medicine.* 9 (10; October): 833–842

Erfo Centrum. 2009. "Facts and Figures about Gene Therapy." URL: http://www.biomedisch.nl/gentherapie_dynamic/gene_therapy_facts_and_figures.php#Number%20of%20trials%201989-2007. Accessed on April 7, 2009.

Fraley, R.T., S.B. Rogers, and R.B. Horsch. 1983. "Use of a chimeric gene to confer antibiotic resistance to plant cells." *Advances in Gene Technology: Molecular Genetics of Plants and Animals. Miami Winter Symposia.* 20:211–221.

Framond, A.J., et al. 1983. "Mini-ti Plasmid and a Chimeric Gene Construct: New Approaches to Plant Gene Vector Construction." *Advances in Gene Technology: Molecular Genetics of Plants and Animals. Miami Winter Symposia.* 20:159–170.

Frederickson, Jodi K. 1997. "He's All Heart . . . and a Little Pig Too: A Look at the FDA Draft Xenotransplant Guideline." *Food and Drug Law Journal.* 52 (4).

"Frontrunners in Securing and Increasing Yield in Crops." 2009. Monsanto press release. Available online. URL: http://monsanto .mediaroom.com/index.php?s=43&item=642&printable?s=43&item= 642&printable. Accessed on April 7, 2009.

Gaxiola, Robert A., et al. 2001. "Drought- and Salt-tolerant Plants Result from Overexpression of the AVP1 H^+-pump."*Proceedings of the National Academy of Sciences U S A.* 98 (20; September 25): 11444–11449.

"Herbicides and Herbicide Resistance in Transgenic Plants." 2009. [Course outline for Horticulture 250 at Purdue University]. Available online: http://www.hort.purdue.edu/hort/courses/HORT250/lecture %2011. Accessed on April 7, 2009.

"History of Plant Breeding." 2009. Available online. URL: http:// www.landfood.ubc.ca/courses/agro/424/resources/AGRO424_ history.pdf. Accessed on April 2, 2009.

Jaenisch, Rudolf, and Beatrice Mintz. 1974. "Simian Virus 40 DNA Sequences in DNA of Healthy Adult Mice Derived from Preimplantation Blastocysts Injected with Viral DNA." *Proceedings of the National Academy of Sciences.* 71 (4; April 1): 1250–1254.

J.E.M. Ag Supply, Inc., DBA Farm Advantage, Inc., et Al. v. Pioneer Hi-bred International, Inc.. 2001. 534 U.S. 124. Available online. URL: http:// caselaw.lp.findlaw.com/scripts/getcase.pl?court=US&vol=534&invol= 124. Accessed on April 6, 2009.

Leader, Benjamin, Quentin J. Baca, and David E. Golan. 2008. "Protein Therapeutics: a Summary and Pharmacological Classification."*Nature Reviews Drug Discovery.* 7 (January): 21–39.

Levy, Marlon F., et al. 2000. "Liver Allotransplantation after Extracorporeal Hepatic Support with Transgenic (hCD55/hCD59) Porcine Livers: Clinical Results and Lack of Pig-to-human Transmission of the Porcine Endogenous Retrovirus." *Transplantation.* 69 (2; January 27): 272–280.

"Macrohistory and World Report." 2009. Available online. URL: http:// www.fsmitha.com/index.html. Accessed on April 1, 2009.

Mawer, Mawer. 2006. *Gregor Mendel: Planting the Seeds of Genetics.* Chicago: Field Museum of Natural History.

Melo, Eduardo O., et al. 2007. "Animal Transgenesis: State of the Art and Applications." *Journal of Applied Genetics.* 48 (1): 47–61.

Murai, Norimoto, et al. 1983. "Phaseolin Gene from Bean Is Expressed after Transfer to Sunflower via Tumor-inducing Plasmid Vectors." *Science*. 222 (4623; November 4):476–482.

Murphy, Denis J. 2007. *People, Plants and Genes: The Story of Crops and Humanity.* New York: Oxford University Press, 246, 331.

National Human Genome Research Institute. 2009. "Cloning." Available online. URL: http://www.genome.gov/25020028#7. Accessed on April 8, 2009.

Schell, J., M., et al. 1983. "Ti Plasmids as Experimental Gene Vectors for Plants." *Advances in Gene Technology: Molecular Genetics of Plants and Animals. Miami Winter Symposia.* 20:191–209.

Sherman, David M. 2002. *Tending Animals in the Global Village: A Guide to International Veterinary Medicine.* New York: Wiley-Blackwell.

Soil Association. 2005. "Flavr Savr Tomato & Gm Tomato Puree: the Failure of the First Gm Foods." Briefing Paper 11/29/2005. Available online. URL: http://www.soilassociation.org/web/sa/saweb.nsf/0/80256cad0046ee0c80256d1f005b0ce5? OpenDocument. Accessed on April 7, 2009.

U.S. Food and Drug Administration. Press Release 94–10. 1994. Available online. URL: http://www.fda.gov/bbs/topics/news/new00482.html. Accessed on April 7, 2009.

Watson, J. D., and F. H. C. Crick. 1953. "A Structure for Deoxyribose Nucleic Acid." *Nature.* 171 (4356; April 25): 737–738.

2

Problems, Controversies, and Solutions

Introduction

Kem Ralph knew that what he was doing was wrong. The Tennessee farmer had agreed to help a neighbor disguise the cotton seeds he intended to plant in 2000. The seeds had actually been saved from a 1999 crop grown with a genetically modified product sold by the Monsanto company. When purchasing the seed originally, the neighbor had agreed not to save seed from the first year's crop to plant the following year, a practice that has been a tradition in farms around the world for centuries. Instead, Ralph agreed to tell Monsanto that the neighbor's seed came from his own crops. He, the neighbor, and many of their friends were angry at Monsanto's policy of restricting the use of their seeds for more than one season. Ralph's plan was discovered, however, and Monsanto decided to sue him for conspiracy to commit mail fraud. He was found guilty, sentenced to eight months in prison, and ordered to pay Monsanto the full value of the seed in question, $165,469, and a penalty of $1.7 million (Robinson 2009.)

Over the past half century, DNA technologies have made possible a host of new opportunities in agriculture, medicine, industry, forensic science, research, and other fields. Scientists now have the ability to change the most fundamental units of which life is made, the molecules that make an artichoke an artichoke, and not a firefly; the molecules that determine how much milk a cow will produce; the molecules that determine whether a child will be healthy or fated to live with a devastating disease.

41

But these new opportunities have also created new challenges for scientists, public officials, and the general public. Some people criticize scientists for "playing God" by manipulating the units of which living organisms are constructed. And in doing so, they raise any number of moral, ethical, philosophical, theological, political, environmental, and social problems. Should plants be genetically altered in order to increase their productivity, if doing so may present a problem for human health and the environment? Should transgenic animals be created for the production of new kinds of foods if both humans and the animals themselves may be at risk from these procedures? Should medical scientists attempt to cure genetic disorders when the cures themselves may be a greater risk than the disease they seek to cure? Should individuals be tested for genetic disorders when test results may be misused by the government or private corporations and individuals may not themselves benefit from the information gained? These are only a few of the issues with which society in the twenty-first century is faced, given the advances in DNA technology.

Forensic Science

All new forms of technology inevitably face challenges from specialists in the field, from legal experts, and from the general public. Will the new technology function as effectively as its proponents claim? Are there risks to the technology that are not yet known? Is the new technology really superior to previous technologies? Questions like these always arise with a new invention or a new technology.

And such was the case with the use of DNA technology in forensic science. When Alec Jeffreys first introduced the possibility of using restriction fragment length polymorphisms (RFLP) for solving crimes in the early 1980s, forensic scientists and law enforcement officials wanted to know if the technology was really (1) as reliable as Jeffreys claimed and (2) superior to previous technologies, such as fingerprint analysis. The first legal challenge to the use of DNA technology in forensic science was posed in a 1989 case in New York state in which one Jose Castro was charged with the murder of a neighbor, Vilma Ponce, and her two-year-old daughter. At the ensuing trial, the prosecution offered evidence obtained from DNA tests that connected Castro with the murder weapon. The defense raised questions about the validity and reliability of the evidence, based in part on the fact that it was such a new technology. The trial judge did not rule on the testing procedure itself, but did bar the evidence because of human errors in the conduct of the testing by Lifecodes laboratory, the company that processed the DNA samples (Aronson 2007).

Similar challenges have occurred on a number of occasions since *People v. Castro*, but they have almost always been based on errors in use of the test, statistical analysis of data, or other related issues, and not on the scientific principles on which the test is based. In fact, by 1994, two scientists who had previously been at odds over the use of DNA fingerprinting in forensic sciences came to the conclusion that "[t]he DNA fingerprinting wars are over ... the public needs to understand that the DNA fingerprinting controversy has been resolved" (Lander and Budowle 1994, 735). One author (Budowle) had long been a supporter of the use of DNA typing in forensic science and had helped establish the FBI's DNA typing program, while the other author (Lander) had long been a critic of the use of this technology for the unambiguous identification of suspects. Over the years, however, the success of the technology had convinced Lander that his concerns were baseless, and the two men agreed that DNA typing was now entirely satisfactory as a valid and reliable method of identifying individuals. This conclusion does not mean, however, that DNA testing in forensic science is no longer a matter of dispute. In fact, the way in which evidence obtained by this technology is utilized remains an issue of concern to a number of specialists in the field and to the general public.

One of the major questions of disagreement is the use of DNA databases. A DNA database is a collection of DNA samples taken from individuals and crime scenes stored at some location. Computer programs are written to compare DNA samples collected at the scene of a crime or taken from individuals arrested at the scene of a crime with samples taken from known criminals. It can also be used to solve so-called *cold cases*, crimes which law enforcement officials have been unable to solve by other means.

The first DNA database in the United States was established by the Federal Bureau of Investigation (FBI) in 1990 as a pilot program serving 14 state and local law enforcement agencies. The software program developed for that undertaking is called the Combined DNA Index System (CODIS). The success of the pilot program led to the passage of the DNA Identification Act of 1994 (now Public Law 103–322) that mandated the creation of a National DNA Index System (NDIS), a collaborative effort among the FBI and state and local law enforcement agencies. As of mid-2009, more than 170 public law enforcement laboratories participated in the NDIS, and CODIS software was being used by more than 40 law enforcement laboratories in over 25 foreign countries (U.S. Department of Justice 2009). The FBI points to the success of the NDIS/CODIS program by noting that the current database contains (as of February 2009) 6,730,749 offender profiles and 255,911 forensic profiles (DNA prints found at crime scenes), and it has produced over 85,600 hits (matches) in more than 84,700 investigations (Federal Bureau of Investigation 2009).

These numbers are impressive, but they should not mask the fact that some controversy remains about the use of DNA databases. As originally conceived, these databases were to contain only DNA samples taken from crime scenes and from individuals convicted of violent crimes, such as murder, rape, and aggravated assault. They were also to be available for the identification of unidentified human remains and in the search for missing persons. Few objections arose about this arrangement. But as databases began to prove their value in obtaining convictions and connecting criminals with specific crimes, some law enforcement officials and politicians suggested expanding the reach of databases. They recommended including DNA samples taken from individuals arrested for or accused of misdemeanors, as well as felonies, whether those individuals were actually convicted of a crime or not. In 2003, for example, President George W. Bush recommended all adults and juveniles who had been arrested for a crime be required to submit a DNA sample (Willing 2003). Many state officials were also enthusiastic about expanding DNA databases. In 2007, for example, New York governor Eliot Spitzer proposed expanding the state's DNA laws to include the collection of DNA samples from anyone convicted of any crime, including misdemeanors such as minor drug offenses, harassment, or unauthorized use of a credit card. He would also have included anyone on probation, under parole supervision, or registered as a sex offender (McGeehan 2007; Lovett 2007). And in New York City, former mayor Rudolph

Giuliani said that he would have "no problem" with fingerprinting all newborn babies in the city (Lambert 1998, B4).

A number of arguments have been put forward for expanding DNA databases. In the first place, law enforcement officers seldom know in advance who is going to commit a crime. The larger a DNA database, the more likely the DNA of a suspect will find a match in such a database. Second, anyone who is innocent of a crime or who has no plans to commit a crime should have no reservations about submitting his or her DNA to a database. Third, a very large database can help to exonerate innocent individuals accused of a crime. All that needs to be done is to compare DNA from a crime scene with the innocent person's databased DNA. Fourth, DNA evidence is a powerful tool not just in identifying criminals, but also in convicting them. Fifth, a large DNA database would probably be a deterrent to crime. A person might be less likely to commit a crime if he or she knew that his or her DNA information was already available to law enforcement officers. Sixth, DNA evidence is often easier to obtain and use than other forms of evidence. It can be obtained, for example, from a single drop of blood or semen, even after it has dried for weeks or months. Seventh, taking of DNA samples can hardly be considered an affront to civil liberties when the information obtained might be useful in preventing and solving terrible crimes.

A number of arguments have been posed against the expansion of DNA databases beyond the minimum absolutely required. Although the test itself is generally regarded as infallible, humans still make mistakes in conducting and interpreting those tests. Innocent people may still be convicted based on faulty DNA evidence. Second, the problem with any database is that once information has been entered, it generally remains there forever. It may not be possible to provide all the controls necessary to keep an individual's DNA information out of the hands of employers, insurance companies, medical companies, and other agencies who have no right to the information. Law enforcement officials must also note that they are not able to keep up with the processing of DNA samples already available to them (the waiting list sometimes runs many months or years), so how will they be able to process even more samples from less dangerous individuals? Perhaps the most troubling of recent recommendations for expanding DNA databases is that most people who are arrested for or accused of a crime are never convicted of that crime. Yet, having to provide DNA samples may well label those individuals as "suspicious," if not actually criminal. Finally, what kind of country would the United States become if every citizen were required to provide a DNA sample, labeling him or her as a potential criminal?

In the United States in 2009, the tendency to include DNA samples from potentially innocent individuals appears to be growing. Fifteen states currently collect DNA samples from people who have been accused of a crime and are awaiting trial, and 16 states require DNA testing of anyone convicted of a misdemeanor. In addition, minors who have been convicted of a crime are required in 35 states to submit DNA samples. And in April 2009, the FBI announced that it would join this trend and begin collecting DNA specimens from individuals who have been arrested or detained for, but not convicted of, a crime. The agency will also begin collecting DNA samples from immigrants who have been detained for any reason. Its argument for this change in policy is that an increased DNA database will result in more convictions of felons (Moore 2009, 1, 19).

The FBI's decision, predictably, caused a furor among civil rights activists who are concerned about privacy rights enshrined in the Fourth Amendment of the U.S. Constitution. As one critic, a lawyer with the American Civil Liberties Union, noted, "What we object to—and what the Constitution prohibits —is the indiscriminate taking of DNA for things like writing an insufficient funds check, shoplifting, drug convictions" (Moore 2009, 19).

Genetically Modified Organisms

Many people, both professional scientists and members of the general public, are very enthusiastic about the potential benefits of transgenic plants and animals in augmenting the nation's and the world's food supply. Probably most important to advocates of genetically modified (GM) foods is the potential for feeding untold millions of people around the world who will otherwise go hungry, become ill, or die from lack of food. Just as important is the fact, they say, that GM foods look and taste at least as good as conventional foods, have the same or better nutritional value, can be stored for longer periods of time, and are easier to ship. Farmers are often enthusiastic about GM crops, because they produce greater yield on the same or smaller amounts of land and require less pesticides and other agrochemicals (San Diego Center for Molecular Agriculture [n.d.]). The biggest problem with GM foods, proponents say, may be that people simply don't know enough about them and much of what they do know may be incorrect. Critics characterize GM foods as "frankenfoods," but they are actually very similar to conventional foods. Finally, advocates argue that the host of issues raised by opponents of GM foods—that they pose a hazard to human health and the environment—are not based on reliable and valid scientific evidence.

For those who oppose GM foods, however, the last point is often the most important; they believe that such foods *do* threaten human health and the natural environment, and they cite a number of specific instances as the basis for this concern. One such concern involves the fact that some individuals are allergic to certain foods, such as nuts, cow's milk, eggs, wheat, and soybeans. These individuals may experience reactions ranging from debilitating to fatal. A possibility exists that a genetically engineered food may contain a gene that might produce an allergic reaction in an unsuspecting consumer. In the 1990s, for example, Pioneer Hi-Bred International, Inc. experimented with the introduction of a gene from the Brazil nut plant (*Bertholletia excelsa*) into soybeans as a way of increasing the nutritional value of the soybeans. As part of its research, Pioneer commissioned a study of possible health effects of this engineered product. The study found that a small number of people who ate the engineered soybean experienced an allergic reaction to the product. Although the soybean was intended for use only as an animal feed, Pioneer decided not to proceed with the product (Genetically Engineered Organisms 2009). Although this story would appear to have a happy ending, critics point to other instances in which GM foods have been implicated with allergic reactions and to the possibilities that such problems may occur in the future with inadequately tested GM products (for a comprehensive review of this argument, see Institute for Responsible Technology 2007).

Another concern involves the development in humans of resistance to certain antibiotics. Researchers commonly add genes for resistance to an antibiotic (such as kanamycin) to an engineered plant or animal because that makes it easy to determine if the genetic transfer has been successful. But what happens if a food intended for consumption by humans carries a gene for antibiotic resistance? If that person later needs to take that antibiotic, the medication may not be effective for the person. Perhaps more important, the gene might then be transferred to the person's offspring, after which it could spread through the general population (BA 2003; for an extended discussion of possible health risks to humans of GM foods, see Seeds of Deception 2009).

Critics of GM foods also point to a number of ways in which such crops can damage the environment. One concern is that genes used in the development of GM crops might escape into the surrounding environment and become integrated into related species, including weeds. In such cases, the weeds would themselves develop the same resistance to herbicides for which the GM crops were developed, and increasingly larger amounts of herbicides would be required to keep them under control. Thus far, there have been relatively few studies to suggest that such a phenomenon, commonly known as *horizontal gene transfer (HGT)*, actually occurs in the field. One such report surfaced in 2002 when a British group reported that weeds growing adjacent to a field of genetically engineered oilseed rape crops in Canada had incorporated herbicide-resistant genes from the cash crop. The scientific advisor for the group, English Nature, warned that the ultimate effect of this case of HGT would be that farmers would have to begin using stronger, more dangerous pesticides to keep the weeds under control (BBC News 2002).

The troubling fact about this discovery is that evidence is now accumulating that HGT is not an uncommon event in nature, as was once thought. An initial hint at the scope of this problem appeared in a paper by a group of biologists at Indiana University led by Jeffrey Palmer. That group found that, "It appears horizontal gene transfer occurs for just about any gene in the plant mitochondrial genome" (IU News Room 2003; Bergthorsson 2003). Five years later, the same researchers reinforced their original findings with another paper reporting on the very widespread occurrence of HGT in nature, not an especially propitious prospect for the wider use of GM crops (Keeling and Palmer 2008).

Critics of genetically modified crops also worry about the effects of those crops on unrelated species in the environment, often referred to as *untargeted species*, because they are not the intended object of the DNA technology being used. One of the best-known studies in this field was conducted by a research team led by Cornell entomologist John Losey. That team removed pollen from genetically engineered corn and spread it on milkweed leaves that are the primary food of monarch butterfly larvae. It found that the larvae who fed on these leaves grew more slowly, ate less, and had a higher rate of mortality than did a control group of monarch larvae. The team concluded that the engineered corn posed a potential threat to the survival of monarch butterflies in areas where the crop was being planted (Friedlander 1999).

The next chapter in this story illustrates some of the problems in dealing with issues surrounding the use of transgenic organisms in agriculture. Six independent research teams followed up on the Losey study, all of them reporting their findings in the October 9, 2001, issue of the *Proceedings of the National Academy of Sciences of the United States of America*. The conclusion drawn from these studies was that the results obtained by Losey's team were not replicated in the field. Instead, the six teams agreed that, for a variety of reasons, the threat posed by the GM corn to monarchs in the field was negligible ("Bt Corn Pollen and Monarch Butterflies." 2001, 11908–11942). The problem is that the follow-up studies received considerably less attention than did the original 1999 studies, so that many people continue to believe the simplified version originally presented and perceive that GM crops are likely to pose a threat to other organisms in the environment (Food and Agriculture Organization. United Nations. 2004, 71). That conclusion is certainly possible, but existing scientific studies suggest that the problem is probably less severe and more complex than it appears at first glance.

The availability of genetically modified crops has transformed the face of agriculture around the world. As of 2009, 92 percent of all the soybeans grown in the United States, 86 percent of all cotton, and 80 percent of all corn came from genetically modified plants (Economic Research Service 2009). These numbers mean that the vast majority of at least some crops being grown come from seeds that can be purchased from one source only: the company that manufactures those seeds. A farmer cannot just go into the local feed-and-seed store and choose from half a dozen seed providers. (The seed provider is likely to be Monsanto, which is now the world's largest supplier of genetically modified agricultural products, with 674 patents for biotechnological products, as of 2008 (Barlett and Steele 2008). One concern, then, is that the world's agriculture system is beginning to fall into the hands of a few very large transnational companies that will determine the character of crops that will be grown everywhere in the world.

A related problem is the degree to which these companies will ensure their control over crops. Most corporations that produce engineered seed require purchasers to sign an agreement that they will not save seeds from one year to be used in the following year (as was the case with Kem Ralph at the beginning of this chapter). These corporations argue that they spend hundreds of millions of dollars in the invention and development of engineered seeds, and it is only fair that they get a reasonable return on their investment. As a result, some companies now have "seed police" that go from place to place looking for violations of agreements that farmers have made with the company. When violations are discovered, as with Ralph, the individuals are likely to be prosecuted and, if convicted, fined, and perhaps imprisoned (Barlett and Steele 2008). Even seemingly innocent individuals may be at risk from the "seed police." In some cases, farmers have been prosecuted because seed from an adjacent plot of land has blown onto their own land, germinated, and grown to maturity. In such a case, those farmers have broken the law because they are growing a patented product (that blew in from the neighbor's field) for which they did not pay or sign an agreement with the manufacturer (The Future of Food 2004; Caruso 2004).

Fewer issues about the use of genetically modified animals have arisen, primarily because no such animals have yet been approved for use as sources of food anywhere in the world. Still, some questions have arisen about other aspects of the genetic engineering of animals. One of the best known issues has surrounded the use of recombinant bovine somatotropin (rbST or rBST or rBGH, for recombinant bovine growth hormone) in dairy cows. Recombinant bovine somatotropin is a hormone implicated in the production of milk in cows. In November 1993, the U.S. Food and Drug Administration (FDA) approved a petition from Monsanto for the use of a genetically engineered form of rBST in dairy cows. The purpose of rBST was to increase milk production in cows. The FDA found that rBST was safe and effective for dairy cows and safe for human consumption. The FDA also found that the use of rBST had no environmental effects and that there are no significant differences between milk from cows that have been treated with rBST and those that have not been treated (U.S. Department of Health and Human Services 1994, 6279).

In spite of this apparent robust endorsement of rBST use in dairy cows, a storm of protest has developed against the Monsanto product. One of the major complaints has been that milk from rBST-treated cows may have a higher-than-normal concentration of insulin-like growth factor-1 (IGF-1), a compound known to be implicated in the development of cancer in humans. Some critics argue that IGF-1 in rBST milk may be 40 times as high as that in untreated milk, vastly increasing the risk of cancer in humans (Kirk 2008). Thus far, studies appear not to have supported this view (International Dairy Foods Association 2007). Nonetheless, a number of major retailers, including Starbucks, Kroger, Walmart, and Safeway have decided not to sell milk from rBST cows.

Concerns over genetically modified foods have reached high levels in some parts of the world, such as Europe and Canada, although public opinion in the United States appears to be somewhat mild on the subject. The Mellman Group has been conducting polls on biotechnology issues for the Pew Charitable Trusts for a number of years and has found that there has been relatively little change in support for (25–27 percent) the greater use of GM foods, while there has been a significant decrease (from 58 percent to 46 percent) in opposition to such foods. A significant factor in these results appears that most Americans seem to know little or nothing about GM foods, with anywhere from 54 to 65 percent of Americans reporting that they have heard, read, or seen nothing about them. When those same individuals are given information about GM foods, acceptance of their role in an American diet increases by about 10 percent among respondents (Mellman Group 2006).

The policies under which genetically modified foods are regulated differ from country to country. In the United States, the general principle that drives regulation is that GM foods have to meet the same standards as conventional foods. A detailed and complex system of regulation has developed to monitor the experimental testing of engineered seeds and the testing of genetically modified foods. But the standards used to approve GM foods are essentially the same as those used for any other new food produced in the United States (Office of Science and Technology Policy 1986; Ramjoué 2007).

Critics of genetically modified crops and foods argue that this policy is too lenient and that more restrictive laws, like those adopted by the European Union, are more appropriate. Having failed thus far to accomplish this goal, U.S. critics have worked instead on legislation requiring that GM foods be labeled. They point out that people have the right to know what they are eating and, in particular, to know if the foods they buy contain genetically modified products (The Campaign 2009). In 2006, Representative Dennis Kucinich (D-OH) introduced HR 5269, "The Genetically Engineered Food Right To Know Act of 2006," requiring the labeling of genetically modified foods. As of 2009, the bill has not yet been considered by the full House of Representatives (H.R. 5269 2006).

Some observers have pointed out problems with the labeling of GM foods. They note that regulations already require food manufacturers to note on labels if a food differs in composition from a conventional food of the same type. But GM foods do not differ in composition from conventional foods, although they do differ in method of production. In addition, deciding which foods would have to be labeled might be an overwhelming problem. Given the complex composition of many foods today in the United States, it is likely that at least one component of a great many foods has been genetically modified in some way or another. After all, more than 80 percent of the corn grown in the United States has been genetically modified in one way or another. Finally, the cost of implementing a labeling program would be far greater than simply printing a new label. It would involve maintaining an extensive and complex system to decide what foods qualify for labeling and what goes on the label. Some authorities have estimated that a labeling program would add on average an additional 10 percent to the cost of labeled foods (McHughen 2008).

As is often the case, individual states have filled a void left by the federal government's failure to adopt legislation on a topic of concern to many people, in this case, genetically modified foods. Forty-two states have adopted laws dealing with one or another aspect of this issue. One category of laws deals with labeling. The state of Alaska, for example, requires the labeling of genetically modified fish and fish products and provides guidelines for the labeling of dairy products from animals that have not been genetically modified in one way or another (Alaska State Legislature 2009, at 17.20.013 and 17.20.040). A number of state laws deal with the problem of crop destruction, an issue that became important when activists in many parts of the country began to take matters into their own hands in dealing with GM crops by destroying crops that were being tested or that had already been modified. For example, a California statute says that

> Any person who willfully and knowingly damages or destroys any field crop product, as specified in Sections 42510 and 52001, that is known by the person to be the subject of testing or a product development program being conducted by, or in conjunction or cooperation with the University of California, the California State University System, or any other federal, state, or local government agency, shall be liable for twice the value of the crop damaged or destroyed .... (California Food and Agricultural Code 2009)

Other laws express their support for biotechnology by providing funds, assistance from state agencies, or other ways of encouraging the development of crop modification within the state. In some cases, those laws are very broad, establishing state agencies to encourage and fund biotechnological research. In other cases, they deal with very limited and specific issues of concern to the state. For example, the New Mexico legislature has adopted a law requiring the state university to "devote an appropriate portion of its funding for the purpose of initiating a program of genetic research and procedures for developing a seed source for faster growing pinon trees suitable to New Mexico's climate" (State of New Mexico 2009).

Xenotransplantation

The use of transgenic animals for xenotransplantation has raised another whole set of questions about the appropriate use of DNA technology in dealing with everyday medical problems. Xenotransplantation is the transfer of cells, tissues, or organs (but usually organs) between two species. In 2009, more than 100,000 people were waiting for a transplanted liver, kidney, pancreas, heart, lung, or intestine (United Network for Organ Sharing

2009). The supply of organs for transplantation from humans is far too small to meet the needs of all patients awaiting a transplant, and, as a result, hundreds of those patients die every year. The use of organs from other mammalian species, especially animals who have been genetically engineered to more closely match human DNA patterns, holds a very significant potential for saving thousands of lives every year.

Xenotransplantation raises some serious technical, ethical, and legal problems, however. Most fundamental are technical questions because, until they are solved, all other issues are essentially irrelevant. As noted in Chapter 1, efforts are now under way to engineer pigs and other animals by introducing genes that reduce the likelihood of rejection of a transplanted organ by a human host's body. This step alone has significantly improved the chance of success of a xenotransplant.

Other technical problems remain, however. One of the most serious is the risk that a transplanted organ will carry with it a virus from the donor animal that then grows and multiplies within the human host body. Such an event would pose a threat not only to the host, but also to his or her progeny, especially since some viruses remain dormant for many years. When the FDA proposed draft guidelines for the use of xenotransplantation, this concern drew the largest number of comments from virologists, other members of the medical and scientific communities, individual citizens, representatives of the American Society of Transplant Physicians and the American College of Cardiology, and commercial sponsors of xenotransplantation clinical trials (U.S. Department of Health and Human Services. Food and Drug Administration. Center for Biologics Evaluation and Research 1999, 2–3).

Another common objection to the use of animals for xenotransplantation is based on the principle that nonhuman animals also have rights. As the group People for the Ethical Treatment of Animals has frequently said, "[a]nimals are not ours to use—for food, clothing, entertainment, or experimentation." In fact, the organization has asked the FDA to ban all xenotransplantation experiments (PETA Medical Center 2009). The most common response to this concern is that humans already keep and slaughter animals for food, clothing, and any number of other purposes, so that using them to save human lives is not really a new use of nonhuman species.

Pharming

For over two decades, pharmaceutical companies have been excited about the possibility of using plants and animals for the production of drugs. As described in Chapter 1, it is possible to insert the genes needed for the synthesis of almost any given protein into the DNA of a cow, goat, sheep, pig, tobacco plant, or other organism, such that that animal or plant becomes a "bioreactor," a living factory for the production of that protein. The desired protein can be harvested from the plant or animal in a number of ways, for example, by collecting an animal's urine, blood, or milk. The protein is then separated from the fluid by chemical means and purified. The final product is normally identical to the protein made by chemical means in the laboratory.

When queried about the potential benefit of this kind of DNA technology, companies almost always point to the reduced cost of the production of drugs. That benefit is not achieved easily, however. For animals, only about one percent of the eggs that receive transformed DNA will, on average, result in a live birth, and only a fraction of those animals will actually express the transcribed gene (Learn.Genetics 2008). Simply producing a single animal with the ability to produce the desired drug, then, can be a hugely expensive challenge. The upside of that story, however, is that once the animal is successfully produced and itself begins to reproduce, the bioreactor system is established. From that point on, other than maintaining the transgenic animals, the desired drug is produced at almost no cost to the pharmaceutical company. According to one estimate, a system of this kind can reduce the cost of some drugs from as much as $1,000 per gram to less than $1 per gram (Meade and Ziomek 1998, 22).

A particularly dramatic example of the potential provided by pharming for drug development was a patent issued in 1999 to the CorpTech company for the production of a tobacco plant carrying a gene for the enzyme glucocerebrosidase, a compound used to treat patients with Gaucher disease. The drug was previously made in an expensive mammalian cell system, which also carried the risk of infections from the cell system. After introduction of the CorpTech recombinant DNA method for producing glucocerebrosidase, the annual cost of treating Gaucher patients dropped dramatically (although annual cost per patient still ranges between $40,000 and $320,000) (Darus and Thye [n.d.], 17).

Glucocerebrosidase is not an uncommon instance of the savings that can be achieved in using transgenic plants for the production of pharmaceuticals. According to one analysis reported in 2002, the average cost of producing a pharmaceutical product with a mammalian cell bioreactor was just over $70 million for 300 kilograms per year. By comparison, the same product could be manufactured at the time for about $60 million using goats, $20 million using tobacco plants, and less than $10 million using corn plants (Pew Initiative on Food and Biotechnology 2002, 5).

For some drug manufacturers, the bottom line is that using mammalian cell systems for producing drugs is just not feasible in the long run. In 2002, 75 percent of all such facilities were being used to produce just four pharmaceutical products. And experts were predicting that by 2010, more than 30 new pharmaceutical products would be competing for those same mammalian cell facilities (Pew Initiative on Food and Biotechnology 2002, 4). Under these circumstances, an alternative method for making pharmaceuticals is badly needed, and transgenic plants appear to be one of the most promising prospects.

Transgenic plants may also hold the promise for the manufacture of novel pharmaceuticals or of products not available by other means. One of the most frequently cited examples is edible vaccines. Injected vaccines have for many decades been a key tool of public health workers in protecting children and adults from potentially serious diseases, such as diphtheria, hepatitis, polio, typhoid fever, typhus, and yellow fever. But injectable vaccines have a number of drawbacks. Many people avoid being vaccinated because of their fear of needles. Delivering needed vaccines to their target populations may be time-consuming, and the vaccines may lose potency while being stored. They may also be destroyed by exposure to high temperatures. And simply because of the way in which they are delivered—by injection—they do not necessarily provide the maximum protection possible.

Some scientists are now hopeful that edible vaccines can be produced that overcome all or most of these obstacles. Genes coding for parts of human pathogens (but not the whole pathogens) are inserted into the DNA of edible plants, such as bananas, potatoes, or rice, and the fruit of those plants is then fed to humans. Inside the human body, the pathogen units stimulate an immune response that is characteristic of vaccines, providing the same or better protection than do injectable vaccines (Langridge 2000, 66–71; Yu 2008, 104–109).

Edible vaccines have their critics, as do most other transgenic plants and animals. The primary concern seems to be that pollen, seed, dust, or plant parts from a transgenic plant might escape from a field and get carried to nearby crops or private gardens. These plant materials could then be absorbed by soil microbes, eaten by farm animals, and/or eventually ingested by humans. The risk posed by this possibility depends to a great extent on the vaccine being produced in the transgenic plant. If, for example, the pharmaceutical being made were a blood thinner, it might pose a threat to the health of humans or animals by whom it was ingested. At this point, more information is needed about the possible combined threat posed by escape of the engineered vaccine and risk posed to human and animal health (Pew Initiative on Food and Biotechnology 2002, 17).

There are currently no specific regulations in the United States for the monitoring of pharmed plants, animals, and products (other than those that apply to all plants, animals, and products). In January 2004, the U. S. Department of Agriculture Animal and Plant Health Inspection Service (APHIS) announced plans to consider the environmental impact of a number of possible revisions to its existing policies on genetically engineered organisms. One of the 10 issues to be considered in this review was "changes that should be considered relative to environmental review of, and permit conditions for, genetically engineered plants that produce pharmaceutical and industrial compounds." The agency listed five possible alternatives in dealing with this issue:

1. No action—continue to allow food and feed crops to be used for the production of pharmaceutical and industrial compounds and to allow field testing under very stringent conditions.
2. Continue to allow food and feed crops to be used for the production of pharmaceutical and industrial compounds. The agency would impose confinement requirements, as appropriate, based on the risk posed by the organism and would consider food safety in setting conditions.

3. Do not allow crops producing substances not intended for food uses to be field tested, that is, these crops could be grown only in contained facilities.
4. Allow field testing only if the crop has no food or feed uses.
5. Allow field testing of food/feed crops producing substances not intended for food uses only if food safety has been addressed (Biotechnology Regulatory Service 2007).

As of mid-2009, these alternatives are being weighed and assessed. The alternative selected will determine U.S. policy on pharmed products for at least the near-term future.

Genetic Testing

The term *genetic testing* actually refers to a number of different procedures, including diagnostic testing, predictive and pre-symptomatic testing, carrier testing, prenatal testing, preimplantation genetic diagnosis, and newborn screening. Diagnostic testing is used when a medical professional has reason to believe that a person has or is likely to develop some form of genetic illness. The testing is done to confirm that diagnosis, improving the ability of a person to choose treatments and other options to prolong his or her life or to make life more comfortable. An example is the test for a disease known as hemochromatosis, a genetic disorder in which a person's body accumulates iron over long periods of time. Excess iron in the body eventually results in a variety of physical problems that may include cardiac dysfunction, cirrhosis, diabetes, and liver cancer. If the problem is diagnosed soon enough, some simple procedures are available for reducing the risk of such problems. One approach is phlebotomy, or bloodletting, in which a certain amount of blood is removed from the body from time to time. Diagnostic tests are now available for the presence of the defective gene that causes hemochromatosis, making it possible for a medical worker to counsel a person about the best way to deal with this disorder. As with all diagnostic genetic tests, the test for hemocrohomatosis can be conducted at any time in a person's life or prior to birth.

Predictive and presymptomatic tests are used with individuals who (usually) have demonstrated no signs of a disease and/or who have reason to believe they may be at risk for a genetic disease. One of the classic conditions for which such tests are used is Huntington's disease (HD), once called Huntington's chorea. Huntington's disease is a genetic disorder that affects the nervous system, producing a loss of coordination, unsteady gait, and jerky body movements, accompanied by a decline in mental abilities and psychiatric and behavioral problems. The condition is incurable.

Perhaps the most troubling aspect of HD is that it typically does not begin to manifest until relatively late in life, after age 40. Predictive and presymptomatic genetic testing makes it possible for a person with a family history of HD to know at any early stage of life whether he or she has the gene responsible for HD and, thus, is likely to develop the disorder. This information makes it possible for a person to plan in whatever way possible for living with and treating the condition.

Carrier testing is used to detect recessive genetic disorders. Some genetic disorders are said to be recessive because they are expressed only when a person inherits a defective gene from both parents. If the person inherits a defective gene from one parent and a normal gene from the other parent, the person does not develop the disease, but is only a carrier for the disease. Knowing that one has one defective copy of the gene and one normal copy of the gene has relatively little significance to an individual; he or she will remain healthy with this genetic background. But this information can be very important to a couple planning to have children. If both parents have one defective copy of the gene (are carriers for the disorder), there is a one-in-four chance that any children they have will receive two copies of the defective gene and will develop the genetic disease. This condition holds for a number of genetic disorders, including cystic fibrosis, Niemann-Pick disease, sickle-cell anemia, spinal muscular atrophy, and Tay-Sachs disease. With the information from carrier genetic testing, a couple can decide whether or not to have children and/or plan how they would care for a child with the designated disorder.

Prenatal testing is often recommended for pregnant women who fall into one of four categories: those who are over the age of 35 (because of increased risk for fetal chromosomal abnormalities); those who have a family history for some genetic disorder, such as cystic fibrosis or Duchenne muscular dystrophy; those whose ethnic or racial background might predetermine a person to a genetic disorder, such as Tay-Sachs disease or sickle-cell anemia; and those who are concerned about the possibility of some other type of genetic disease. Prenatal testing is usually conducted by amniocentesis or chorionic villus sampling (CVS). In either case, a small sample of fetal material is removed and examined for genetic or chromosomal abnormalities. The results of these tests help couples and their health providers to consider birth and care options for the unborn fetus and, later, the newborn child.

Preimplantation genetic diagnosis (PGD) is a relatively new type of genetic testing. Women who hope to become pregnant can have eggs that have been fertilized *in vitro* (outside the human body) examined for possible genetic or chromosomal defects before they are implanted. Before the availability of PGD, the only way to test for such abnormalities was to proceed with the implantation and then use amniocentesis or CVS at some later date to look for possible fetal defects.

Newborn screening is a term that refers to a testing program used with every newborn child, whether any reason exists for concern about possible genetic disorders or not. A screening program is used to identify certain somewhat common disorders that can and should be treated as soon as the child is born, phenylketonuria (PKU) being perhaps the best known example. Left untreated, PKU can be fatal. But it can be treated very easily by providing a newborn child with a specialized diet, and the child can then live a normal life as long as he or she continues on that diet. All states now require some kind of newborn screening panel in which tests for a minimum of 20 disorders are conducted (National Newborn Screening and Genetic Resource Center 2009).

Risks and Benefits

There is probably no question that genetic testing can have very significant benefits for parents and children. The detection of PKU is perhaps the best single example. If the condition is detected before a child is born, dietary adjustments can be made in her or his life that make a potential life-threatening condition into an easily controlled disorder. Another example is the genetic condition known as familial adenomatous polyposis (FAP). Individuals with this condition develop numerous benign polyps (growths) on their colon, often beginning as early as their teenage years. Over time, these polyps tend to become cancerous, resulting in a life-threatening condition. If an individual knows in advance that she or he carries the genes for FAP, steps can be detected to reduce or eliminate the severity of the disorder (Genetics Home Reference 2009).

Still, a number of concerns exist about the use of genetic testing for pregnant women, newborn children, and adults. In the first place, the possibility of error always exists in genetic tests, as it does in any human endeavor. Chemicals may not react in the manner expected (and the manner they do 99 percent of the time), humans may make errors in performing a test, or they may make errors in reading and interpreting the test. Such errors can have profound results. A woman told that her unborn child carries the gene for cystic fibrosis, for example, might decide not to continue the pregnancy. But what if that result were erroneous? The woman's decision would have been based on incorrect information and would have resulted in an act to which she would otherwise never agree.

A second problem concerns the usefulness of information obtained from a genetic test. The number of monogenic disorders (diseases caused by an error in a single gene) is relatively small, and even monogenic disorders can be complex and unpredictable. For example, more than 300 different mutations are possible in the gene that causes cystic fibrosis, and each mutation produces a disease of differing severity and characteristics. To discover that one is carrying an erroneous gene for cystic fibrosis provides, therefore, only minimal information for its possible treatment (Andrews, et al. 1994, 62).

How valuable is the information, then, that one is carrying one or more genes for some incurable disorder, such as some types of breast cancer, cystic fibrosis, or Huntington's disease? Granted, such information allows a person to make lifestyle choices, financial decisions, end-of-life choices, family planning, and other arrangements about one's future. But at what emotional cost? Imagine the turmoil experienced by a 20-year-old man who has just tested positive for Huntington's disease, which may not manifest for 20 years or more. The man knows that his disease, if and when it develops, is incurable, but he also now has the option to decide whether to have a family and how to plan for his future. Is he better or worse off for having had the genetic test? Each individual will answer that question differently.

Another common concern about genetic testing arises over questions of privacy and confidentiality. There is probably nothing more obviously a person's own private business than her or his genetic information. How many people are willing to share with outsiders basic data about their own genome? Genetic testing companies usually take extensive precautions about the genetic data they collect from people. Yet, those data can easily fall into the hands of outsiders, most commonly, insurance agencies. In order to issue a health or life insurance policy to an individual, a company may want to know (understandably) everything possible about a person's health condition. Insurance companies are typically not eager to issue policies to people who have fatal illness or are at risk for serious health problems. If they know a person has had genetic tests, they will probably want to know the results of those tests, and not insure those with identified long-term serious health problems. Is this a legitimate use of information obtained from genetic tests?

One of the most recent issues related to genetic testing has been the use of testing kits designed for consumer use. Such kits are sold on the Internet or through other outlets directly to consumers. They usually have the consumer take a DNA sample (for example, by swabbing the inner surface of one's cheek) and send that sample to the sponsoring company for analysis. These kits commonly carry clear and specific warnings to the effect that they are not designed for making specific diagnoses. In 2006, the U.S. General Accountability Office (GAO) conducted a study to assess four testing kits purchased on the Internet. They submitted cheek swabs and complete medical histories from 14 fictitious individuals to the four companies from whom they had purchased the kits. (All 14 swabs were actually obtained from a nine-month-old female). The GAO study found, first of all, that all four kits they ordered online did contain adequate warning statements, such as "[This is] not a genetic test for disease or predisposition to disease, nor does it determine a medical condition," and "Please note that this screening is not a test for inherited disorders" (U.S. Government Accountability Office 2006, 8). But the study also found that all four companies reported that the 14 fictitious individuals were at risk for developing one or more genetic disorders, including cancer, osteoporosis, type 2 diabetes, high blood pressure, and heart disease. The companies then recommended a variety of treatments for the prevention and/or treatment of these conditions. In most cases, those treatments involved the use of dietary supplements, which the companies offered at costs of up to 40 times the expense of the same supplements available at local stores. The GAO concluded from its study that "all the tests GAO purchased mislead consumers by making predictions that are medically unproven and so ambiguous that they do not provide meaningful information to consumers" (U.S. Government Accountability Office 2006, [i]).

Genetic Counseling

The availability of genetic testing has spawned the development of a new career field: genetic counseling. The purpose of genetic counseling is to provide professional assistance to individuals and couples who are contemplating having genetic tests for one or more disorders or who have already had such tests and wish to better understand what those tests mean and what their possible implications are. Professional genetic counselors typically have earned a master's degree in medical genetics and counseling and have passed a certification examination administered by the American Board of Genetic Counseling (American Board of Genetic Counseling 2009).

Some people choose to meet with genetic counselors before they become pregnant; some, after they have become pregnant, but before they have had any genetic tests performed; and some, after tests have been completed and results are available. In each case, a counselor can review the nature of any one or more genetic tests, the kind of information produced by the tests, the relative seriousness of conditions indicated by positive test results, the likelihood that a positive test result indicates that a child will actually develop a given condition, pre-birth options for parents of the fetus, post-birth treatments and maintenance procedures that may be available, and other relevant issues that parents may have to deal with if a child carries a gene or chromosome associated with some specific disorder (*Making Sense of Your Genes: A Guide to Genetic Counseling* 2008).

Regulation

Until recently, only relatively modest federal legislation existed dealing with genetic testing. Standards for clinical laboratories in general has long been under the purview of the Food and

Drug Administration, but there has been no specific regulation of genetic testing on a national level. That situation changed in 2008 when the U.S. Congress passed and President George W. Bush signed the Genetic Information Nondiscrimination Act, which prevents employers and health insurance companies from using information obtained from genetic tests to discriminate against individuals. The bill received virtually unanimous support in the Congress, passing the Senate by a vote of 96 to 0 and the House by a vote of 420 to 3, with nine abstentions. Proponents of gene testing regulation were pleased with these results. Senator Ted Kennedy (D-MA), for example, called the legislation "the first major new civil rights bill of the new century" ("Kennedy in Support of Genetic Information Nondiscrimination Bill" 2008).

In spite of the high praise and widespread support for the bill, some criticism of the legislation remains. One concern is that the bill is not as inclusive as some observers would like. For example, it does not protect people from discrimination when applying for life insurance or for long-term care or disability insurance (Keim 2008). Employer organizations such as the National Association of Manufacturers, National Retail Federation, Society for Human Resource Management, and U. S. Chamber of Commerce argue that provisions of the bill may be too onerous for many companies, requiring them to offer insurance to many people who have serious health problems. They also are concerned about possible conflicts between the new federal law and the many different state laws dealing with genetic testing (Fletcher 2007).

Most states have also passed legislation protecting the privacy of individuals who have had genetic tests. In most cases, those laws require that an individual give specific and informed consent for the use of information obtained from genetic testing to any other person or agency. These laws usually also prohibit a third party from requesting or requiring that an individual have a genetic test. A number of states also have laws that define DNA samples and/or the information they contain as personal property, allowing them to be considered within that category for legal purposes. Four states—Delaware, Nevada, New Mexico, and Oregon—specifically limit information from genetic tests to individuals only (National Conference of State Legislatures 2008). One of the simplest laws is that of Hawaii, which has three provisions: (1) that genetic information cannot be used in making any decision as to whether a person is eligible for health insurance or the type of insurance that can be offered; (2) an insurance company cannot request or require that an applicant for insurance have any kind of genetic test; and (3) information obtained from genetic testing cannot be given to any other person or organization without an individual's written approval.

Gene Therapy

As with other forms of DNA technology, the primary issue relating to human gene therapy (HGT) with which practitioners and the general public have to deal is technical problems. The early successes of HGT experienced by W. French Anderson and other researchers in the late 1980s and early 1990s (see Chapter 1) inspired a great deal of hope among researchers and patients with potentially fatal or severely disabling disorders and their families. A few deaths and other disastrous results, however, clearly highlighted the technical problems involved in treating genetic disorders as readily and successfully as early pioneers may have hoped. A number of companies working in the field of HGT simply gave up on the technology, while others have struggled through difficult times because of diminished hopes for the technology (Timmerman 2009).

Should technical problems involved in the use of gene therapy be solved and the technology become more generally available to individuals with severe genetic disorders, it seems possible that considerable public support for the procedure will develop. It seems somewhat difficult to imagine, absent technical risks for patients, how an individual would refuse a treatment that would save his or her life or vastly diminish the medical problems required simply to stay alive. Even then, however, some ethical questions remain about the use of gene therapy for the treatment of genetic disorders. Perhaps the most frequently mentioned of these issues is the so-called "slippery slope" argument. The slippery slope argument says, in general, that allowing some action to occur in the present, even if that action seems harmless and benign, may set into action a series of other events with far less benign consequences. That is, the first step at the top of a slippery slope may not pose any problems whatsoever, but once that step is taken, there may not be any way to avoid a whole host of less pleasant circumstances, such as sliding all the way to the bottom of the slope.

In dealing with gene therapy, the slippery slope issue usually focuses on the distinction between therapeutic and enhancement (cosmetic) treatments. A therapeutic treatment is one designed to cure a disease, such as cystic fibrosis or sickle-cell anemia. An enhancement treatment is one intended to improve one's overall physical or mental condition, that is, making a person better looking or more intelligent (or, conceivably, more personable, more aggressive, more intuitive, more friendly, more . . . the list goes on). It might appear that the two kinds of treatments are very different from each other. But such is not necessarily the case. How would a person with mild allergies be classified? Those allergies can probably be controlled by avoiding the allergens responsible for one's condition. But if a genetic component of the allergic reaction could be found, would not gene therapy be a better long-term solution for the person's problem? Or what about a child who fails to grow normally (whatever "normal" might mean in this case)? If a defective gene responsible for this problem could be found, should not the person be offered the option of having gene therapy to "cure" the problem?

In any case, many people clearly feel that certain physical characteristics can be a handicap in life and are willing to pay to have those characteristics altered. In 2007, nearly 11.7 million cosmetic surgical procedures were conducted in the United States, ranging in price from relatively simple laser hair removal (at an average cost of $387) to lower body lifts (at an average cost of $8,043). The most common cosmetic procedures conducted were botox injection (329,519 cases), laser hair removal (185,684 cases), microdermabrasion (85,910 cases), hyaluronic acid treatment (84,184 cases), liposuction (57,980 cases), and eyelid surgery (32,564 cases) ("Cosmetic Plastic Surgery Research" 2009). With this level of interest in cosmetic surgery, some authorities are concerned that gene therapy will eventually become used for dealing with almost any "defect" that a person might imagine. David King, a former molecular biologist and director of the watchdog group Human Genetics Alert, has observed that:

> The fact is that science is not merely changing the barrier between disease and "'normality": it is changing our whole conception of what it is to be human. There is an increasing reluctance to accept any predetermined

limits of our biology and a tendency to regard human body as being perfectible .... When gene therapy becomes familiar, we will start to see our genes as less finalised. People are increasingly coming to feel that they are no longer stuck with the body they were born with. The definition of the human condition has now become a function of technology .... (King 1997; see also Harris and Chan 2008, 338–339)

King has also expressed concern about another potential use of gene therapy. In general, gene therapy can be conducted along two lines: using somatic cells as the recipients of engineered genes, or using germ line cells. Somatic cells are any cells other than those that make up the reproductive system. Any change made in a somatic cell affects the organism into which it is introduced, but not its descendants. By contrast, genetic information contained in germ line cells (sperm and eggs, for example) is passed on from generation to generation, so that any modification made in such cells becomes part of the genome of succeeding generations. The only medical reason for modifying germ line cells is to stop the transmission of a genetic disorder from one generation to the next and, perhaps, to eliminate the disorder from the human race entirely. Thus far, for a variety of reasons, there has been virtually no research on the genetic modification of germ line cells. However, the medical potential of treating genetic disorders by germ line modification has been discussed seriously by researchers for at least a decade. Some of those researchers believe that the use of germ line, rather than somatic, cells is "an ideal form of gene therapy" (Fox 1998, 407). That may well be the case in the pursuit of methods for ending genetic disorders, but is it as acceptable for use in enhancement therapy?

Finally, the most powerful argument against gene therapy among some people is that it represents an effort on behalf of scientists to modify the very nature of human beings. For these individuals, gene therapy is nothing other than "playing God," taking over responsibilities that are not allowed to humans. This argument quickly becomes very complex, as it is perfectly obvious that humans act as if they were "playing God" all the time. When a doctor acts to cure a disease, for example, he or she appears to be interrupting a fate that has been decreed by a higher power, if one does believe in such a power. In the case of genetic therapy, the problem is made difficult because some ethicists argue that somatic genetic therapy is acceptable because it is a curative act, while germ line genetic therapy is forbidden because it alters the very nature of human beings ("Responsa in a Moment: Genetic Engineering" 2009).

The ethical debate over gene therapy appears to have reached its peak in the 1990s when breakthroughs in the technology were anticipated to occur at any moment. In fact, progress in the field has been much slower than many experts believed at the end of the last century, and there appears to be less concern about moral and ethical issues today and more interest in a variety of technical problems. Still, an optimistic view is that the potential for human gene therapy will one day be realized, and, at that point, the moral and ethical dilemmas outlined here will once again come to the forefront (Szebik and Glass 2001, 32–38).

Regulations

The U.S. Food and Drug Administration is the lead federal agency in the regulation of human gene therapy research and development activities. In October 1993, it issued a general directive defining in detail its statutory authority for regulating various aspects of the development, testing, and use of new gene therapy technology (U.S. Department of Health and Human Services. Food and Drug Administration 1993, 53248–53251). Since that time, the agency has continued to issue directives dealing with specialized aspects of research in the area of human gene therapy. Some examples of the topics covered in these directives are *Draft Guidance for Industry: Potency Tests for Cellular and Gene Therapy Products, Guidance for Industry: Gene Therapy Clinical Trials— Observing Subjects for Delayed Adverse Events, Guidance for Industry: Guidance for Human Somatic Cell Therapy and Gene Therapy,* and *Guidance for Industry: Eligibility Determination for Donors of Human Cells, Tissues, and Cellular and Tissue-Based Products* (U.S. Food and Drug Administration 2009). Of special interest to the general reader are occasional articles on the topic for the general public (see, for example, Thompson 2000, 19–24 and Witten 2007).

Cloning

The term *cloning* refers to any process by which an exact copy of a gene, cell, organism, or other entity is made. Cloning is a very common natural process. When a bacterial cell reproduces asexually, for example, it splits into two genetically identical daughter cells. Those daughter cells then split again, into four new, but genetically identical cells. The four split into eight, and so on. The same process occurs at the early stages of growth in all plants and animals. A fertilized human egg, for example, divides into two daughter cells, each of which is identical to the original cell, a process that is repeated many times over. This process, by which a group of genetically identical cells are produced from a single parent cell, is called *cell cloning*.

Cell cloning that occurs in plants is also known as *vegetative reproduction*. A number of plant species reproduce by this method. Such plants may reproduce asexually by producing buds on leaves, sending out runners from stems, or producing new structures on roots. For centuries, horticulturists have used vegetative reproduction for the cultivation of a number of crops by rooting cuttings of a plant or grafting tissue from one plant to another.

The research of Paul Berg, Stanley Cohen, Herbert Boyer, and their colleagues in the early 1970s made possible another type of cloning, *molecular cloning*. By using recombinant DNA technology, it is now possible to produce clones of whole molecules or fragments of molecules. The molecular segment to be cloned is inserted into a plasmid, which is then inserted into a bacterium. As the bacterium reproduces and carries out its normal life functions, it expresses not only its own DNA, but also any information stored in the DNA fragment that has been added to its own genome. If the fragment added is a gene (as it almost always is), the bacterium produces the protein for which the gene codes. The natural cloning of the original bacterial cell results in countless numbers of daughters, all of which are identical to the original cell, and all of which produce the protein for which the mother cell was engineered. This procedure has been employed in the production of transgenic plants and animals and in pharming, discussed earlier in this chapter, resulting in the availability of a host of new agricultural, pharmaceutical, and industrial products.

Two other forms of cloning, *therapeutic cloning* and *reproductive cloning*, have become the focus of intense debate among politicians, scientists, theologians, and the general public. The two types of cloning have very different objectives. The purpose of therapeutic cloning is to generate and then harvest specific cells that can be used in treating medical disorders. The goal of reproductive cloning, on the other hand, is to produce an entirely new organism. Both types of cloning make use of a common procedure, *somatic cell nuclear transfer* (SCNT). The first step in SCNT is to remove the nucleus of an ovum (egg cell) which, in a sense, makes the enucleated (without a nucleus) cell a factory ready to make proteins, without any directions (DNA) as to what proteins to make. Next, the nucleus of a somatic cell is removed and inserted into the enucleated ovum, a process known as *transfection*. The cell factory now has directions as to the products (proteins) it is to make. The hybrid cell is then stimulated by some means to begin dividing. As it divides, it produces clones, all of which are genetically identical to the somatic cell from which the nucleus was originally taken. If that process is allowed to proceed without outside control, a new organism develops that is genetically identical to the one from which the somatic DNA originally came. If that somatic cell came from a frog, the animal that develops will be a frog; if a sheep, it will be a sheep; if a human, it will be a human. Somatic cell nuclear transfer used for this procedure, thus, is a method of reproductive cloning, since it results in the formation of a complete adult organism identical to the parent organism from which the transferred nucleus was taken.

To say that SCNT results in reproductive cloning is not precisely true. While the vast amount of genetic information in a eukaryotic organism is stored in nuclear DNA (DNA in the nucleus of its cells), some small amount of genetic information is also found in mitochondrial DNA. Mitochondrial DNA is DNA found in the mitochondria located in the cytoplasm of a cell. It is closely related to ancestral prokaryotes from which eukaryotes originally evolved. In a SCNT procedure, the vast amount of the genetic information contained in clones will be that from the nuclear DNA transferred from the somatic cell donor. That, of course, explains why adult clones produced by SCNT procedures look so much like the parent organism. But some small amount of genetic information in the clones derives from mitochondrial DNA residing in the original enucleated ovum. One of the fundamental questions still unanswered in SCNT research is how nuclear DNA from the somatic cell donor and mitochondrial DNA from the host egg interact, and what gross effects such interactions will have on the health and well-being of clones produced by SCNT (Committee on Science,

Engineering, and Public Policy. Board on Life Sciences 2002, 47–48).

Therapeutic cloning very much resembles reproductive cloning in that DNA from a somatic donor cell is transfected into an enucleated host ovum. The host cell then begins to replicate until it has reached the blastocyst stage of development. A blastocyst is an early stage embryo that develops in humans about four to five days after fertilization of the ovum. It consists of fewer than 150 cells. The portion of the blastocyst of interest to scientists is a group of cells on the inner wall of the blastocyst known as *inner cell mass*. These cells are known as *embryonic stem cells*. Embryonic stem cells are of interest to researchers because of two properties: (1) under the right conditions, they are capable of reproducing as undifferentiated cells virtually forever; and (2) under other conditions, they can be stimulated to differentiate into any one of the more than 200 specialized cells (blood cells, nerve cells, muscle cells, etc.) present in the human body.

Controversies about Cloning

Some of the most robust debates over DNA technology center on the use of cloning technologies. Disagreements over the use of molecular cloning for the production of transgenic plants and animals have been discussed above. At least as controversial has been the use of somatic cell nuclear transfer for therapeutic and reproductive cloning.

This debate has been somewhat clouded by the confusion between therapeutic and reproductive cloning, particularly when applied to humans. Probably the most common generalization that can be made is that the vast majority of individuals and organizations oppose the reproductive cloning of humans, even though those same organizations and individuals may then have very different views about therapeutic cloning.

Human Reproductive Cloning

That having been said, some arguments have been put forward in favor of human reproductive cloning. For example, the technology provides a way for women and couples who are otherwise unable to have children by conventional means to bear a child. Such individuals would include the relatively small number of men who do not produce sperm and women who do not produce eggs, as well as gay men who choose not to have children with genes from a female and lesbians who choose not to have a male sperm donor (Strong 2005, 654–658). Some proponents of reproductive cloning have also suggested that the technology provides a method for replacing a dead child very much missed by its parents (Connor 2008). Whether one agrees with these arguments or regards them as absurd or outrageous, many advocates of human reproductive cloning argue that the fundamental point may be, after all, that cloning is a technology that individuals should have the right to choose or not, as their own conscience dictates (see, for example, Pearson 2006, 657–660).

Some observers have gone even further and claimed that reproductive cloning might actually be preferable to conventional means of reproduction. One of the world's leading bioethicists, Joseph Fletcher, has written that cloning would not be a dehumanizing activity, as claimed by its opponents, but a good choice from the standpoint of both individuals involved and the general society. At one point, he has argued that

> Good reasons in general for cloning are that it avoids genetic diseases, bypasses sterility, predetermines an individual's gender, and preserves family likenesses. It wastes time to argue over whether we should do it or not; the real moral question is when and why. (Fletcher 1974, 154)

Most people do not find the arguments in favor of reproductive cloning to be persuasive. Many polls in the United States, Europe, Australia, and other countries of the world have found large majorities of respondents in opposition to the practice (Center for Genetics and Society 2003). One of the most recent of those polls, conducted by the Virginia Commonwealth University (VCU) Center for Public Policy in 2008, found that 78 percent of respondents were opposed to research on human reproductive cloning ("VCU Life Sciences Survey" 2008, 2). These results, along with those of other polls, can be somewhat difficult to interpret since respondents are often not very knowledgeable about the differences between reproductive and therapeutic cloning. In the 2008 VCU poll, for example, a third (33 percent) of all respondents said that they were "not at all clear" as to the difference between reproductive and therapeutic cloning, 31 percent were "not very clear," and only 34 percent were "somewhat clear" or "very clear" (8 percent) ("VCU Life Sciences Survey" 2008, 16).

When individuals are asked their view about human reproductive cloning, they often respond with an off-the-cuff emotional response that one observer has described as "a group of
responses based on yuk, it's unnatural, it's against one's conscience, it's intuitively repellent, it's playing God, it's hubris"
(Gillon 1999, 3–4). Although clearly heartfelt and sincere, this
response is perhaps difficult to assess rationally. As the observer
quoted here observes,

> The trouble is that these gut responses may be morally
> admirable, but they may also be morally wrong, even
> morally atrocious, and on their own such gut responses
> do not enable us to distinguish the admirable from the
> atrocious. (Gillon 1999, 4)

Other arguments are, perhaps, easier to consider and assess. For
example, some critics of human reproductive cloning say that the
practice diminishes the value of the individuality and unique-
ness of the clone. A person born as a result of this process might
be thought of as a "manufactured product" rather than a special
human being, deserving of his or her own unique value. Part of
the problem might also be that the cloned person would forever
live in the shadow of the person from whom he or she was
cloned, further degrading the clone's individuality. Finally, using
human reproductive cloning for some possibly valid reasons
might be another example of taking a step down that "slippery
slope" in which the technology might then be used in the future
for many less salubrious applications (Center for Genetics and
Society 2006a).

At the moment, the arguments for and against human reproductive cloning are all somewhat theoretical. The problem is that
current technology has not yet developed to a point where the
procedure is safe enough to think of using it with humans, even
on an experimental basis. In 2002, a special panel on Scientific
and Medical Aspects of Human Cloning of the National Academy of Sciences published a report on the scientific and medical
aspects of human reproductive cloning. The committee did not
consider ethical and moral questions in its deliberations, simply
what the current status of research in the field was. It concluded
that the technology was still in a very primitive stage and was
not safe to use with humans, even on an experimental basis.
One of its findings was that

Data on the reproductive cloning of animals through the use of nuclear transplantation technology demonstrate that only a small percentage of attempts are successful; that many of the clones die during gestation, even at late stages; that newborn clones are often abnormal or die; and that the procedures may carry serious risks for the mother. In addition, because of the large number of eggs needed for such experiments, many more women would be exposed to the risks inherent in egg donation for a single cloning attempt than for the reproduction of a child by the presently used in vitro fertilization (IVF) techniques. These medical and scientific findings lead us to conclude that the procedures are now unsafe for humans. (Committee on Science, Engineering, and Public Policy. Board on Life Sciences 2002, 93, Appendix B)

Therapeutic Cloning

In contrast to public opinion about human reproductive cloning, opinions about therapeutic cloning among the general public are more closely divided. In the 2008 VCU poll cited above, 52 percent of respondents said they favored the use of cloning technology for therapeutic purposes (23 percent "strongly" and 29 percent "somewhat strongly"), while 45 percent opposed using the technology. A somewhat more recent poll, conducted by the *Washington Post* in March 2009, found very similar results, with 59 percent of all respondents in favor of therapeutic cloning and 35 percent opposed (Cohen 2009).

In the case of therapeutic cloning, proponents and opponents of the practice commonly cite a single major argument in support of their position. For those who favor the practice, that argument is that the embryonic stem cells produced by therapeutic cloning may be useful in curing any number of diseases and disorders, including Parkinson's disease, amyotrophic lateral sclerosis, spinal cord injury, burns, heart disease, diabetes, and arthritis (Stem Cell Information 2009). While it might be unfortunate to have to destroy early stage embryos to obtain the stem cells needed for this research, the price is a small one to pay for finding cures for such diseases. Besides, the procedure used to make stem cells for therapeutic research is already in use for in vitro fertilization, so scientists are not breaking a new ethical barrier in wanting to use cloning to find the cure for disease.

Finally, many people believe that an early stage embryo has a different moral value than does a later stage embryo, a fetus, or a newborn child.

For opponents of therapeutic cloning, the primary concern usually is that the early stage embryo has the potential to become a living human being. There is no sharp line of demarcation between "not-human" and "human," only a slow progression from primitive human to a living human. It makes no sense, critics say, to destroy a life (that of the early stage embryo) in order to save another life (that of someone with a life-threatening disease). Each life is equally valuable. Furthermore, the argument that therapeutic cloning is used for in vitro fertilization is inappropriate, since critics believe that *that* application is also morally wrong. Finally, a number of alternatives to the use of embryonic stem cells for therapeutic cloning now appear to be available, so the use of that technology may no longer be justified at all (Center for Genetics and Society 2006b).

Critics of using embryonic stem cells point to an important breakthrough that occurred in 2007. Two research teams, one led by Shinya Yamanaka of Kyoto University, and the other by Junying Yu, working at the University of Wisconsin-Madison, found a way to convert skin cells into cells with the same characteristics as those of embryonic skin cells. If these results can be confirmed, the discovery provides a new method for using stem cells that does not require that cells with the potential for life be sacrificed. In such a case, scientific researchers may have found a solution to a difficult problem that politicians have thus far been unable to solve.

Regulations

Beginning in 1997, the U.S. Congress has made a number of attempts to pass legislation dealing with reproductive cloning (Center for Genetics and Society 2009; Religious Tolerance 2007). In some cases, bills have failed to pass either house, while in other cases, a bill has passed the House of Representatives, but failed in the Senate. The primary reason for the lack of federal legislation thus far appears to be an inability to distinguish reproductive from therapeutic cloning. A bill that would ban reproductive cloning alone would, based on previous experience, appear to have a good chance of passing both houses and being signed by a president. But enough people in one or both chambers feel strongly enough that therapeutic cloning should also be banned, that all bills so far have also included that provision, which makes them unacceptable to a majority of the Senate.

An important breakthrough in this regard occurred in March 2009 when President Barack Obama signed an executive order allowing the use of federal funds for stem cell research, a step that had previously been prohibited by former President George W. Bush. In addition, Obama expanded the sources from which stem cells could be taken for research purposes, although those sources were limited to fertility clinics which would otherwise have disposed of the cells.

By contrast, a number of states have adopted some type of legislation banning or regulating cloning. The first such act was adopted by the California legislature in 1997. It banned reproductive cloning or cloning for the purpose of inducing artificial pregnancy, but it permitted cloning for research purposes. Five other states—Connecticut, Maryland, Massachusetts, New Jersey, and Rhode Island—have passed similar bills, banning reproductive cloning but permitting therapeutic cloning, while six states—Arkansas, Indiana, Iowa, Michigan, North Dakota, and South Dakota—have passed laws explicitly banning both kinds of cloning. Arizona and Missouri have adopted legislation prohibiting the use of public funds for cloning, and one state law, in Virginia, is sufficiently ambiguous to make it unclear what the legislature's position was on therapeutic cloning (although reproductive cloning is clearly banned) (National Conference of State Legislatures 2009).

Conclusion

DNA technology has made possible a dazzling array of new procedures and products to aid in fighting crime, to augment and improve the world's food supply, to help medical sciences have a better understanding of the nature of genetic illnesses and to provide pharmaceuticals to use against those and other kinds of disease, to find ways of curing or even eliminating some genetic disorders, and to produce exact copies of animals that can then be used in a host of beneficial ways. None of these technologies has been quickly and easily adopted, however, as each brings with it certain risks to human health and/or the environment and, in many cases, raises profound social, political, and ethical questions. As is usually the case with progress in science, the great challenge is to find ways of gaining the greatest benefit of each new technology while reducing to the greatest extent possible any serious risks posed to the natural and human world.

References

109th Congress. 2nd Session. 2006. "H.R. 5269." Available online. URL: http://www.govtrack.us/congress/bill.xpd?bill=h109-5269. Accessed on March 14, 2009.

Alaska State Legislature. 2009. Infobases. Available online. URL: http://www.legis.state.ak.us/basis/folio.asp. Accessed on April 17, 2009.

American Board of Genetic Counseling. 2009. Available online. URL: http://abgc.iamonline.com/english/view.asp?x=1. Accessed on April 15, 2009.

Andrews, Lori B., et al, eds. 1994. *Assessing Genetic Risks*. Washington, D.C.: National Academy Press.

Aronson, Jay D. 2007. *Genetic Witness: Science, Law, and Controversy in the Making of DNA Profiling*. New Brunswick, NJ: Rutgers University Press, 2007, Chapter 4.

BA [British Science Association]. 2003. "Open Meeting: GM Foods: Are They Safe to Eat?" Available online. URL: http://www .britishscienceassociation.org/NR/rdonlyres/75AB1B69-93E2-46A9-9736-C7BA5E5AF3CD/0/FinalReport.pdf. Accessed on April 11, 2009.

Barlett, Donald L., and James B. Steele. 2008. "Monsanto's Harvest of Fear." *Vanity Fair*. May 2008. Available online. URL: http://www .vanityfair.com/politics/features/2008/05/monsanto200805. Accessed on April 12, 2009.

BBC News. 2002. "Rogue GM Plant Warning." Available online. URL: http://news.bbc.co.uk/1/hi/sci/tech/1800322.stm. Accessed on April 11, 2009.

Bergthorsson, Ulfar, et al. 2003. "Widespread Horizontal Transfer of Mitochondrial Genes in Flowering Plants." *Nature*. 424 (July 10):197–201.

Biotechnology Regulatory Service. U.S. Department of Agriculture. 2007. *Introduction of Genetically Engineered Organisms*. Washington, D.C.: U.S. Department of Agriculture.

"Bt Corn Pollen and Monarch Butterflies." 2001. *Proceedings of the National Academy of Sciences of the United States of America*. 98 (21; October 9): 11908–11942.

California Food and Agricultural Code. Section 52100. 2009. Available online. URL: http://www.aroundthecapitol.com/code/fac/52001-53000/52100. Accessed on April 17, 2009.

Caruso, Denise. 2004. "New 'Future of Food' Movie is a Hit!" Organic Consumers Association. Available online. URL:http://www.organic consumers.org/ge/future-of-food.cfm Accessed on April 12, 2009.

Center for Genetics and Society. 2003. "CGS Summary of Public Opinion Polls." Available online. URL: http://www.geneticsandsociety.org/article.php?id=401. Accessed on April 16, 2009.

Center for Genetics and Society. 2006a. "Reproductive Cloning Arguments Pro and Con." Available online. URL: http://www.genetics andsociety.org/article.php?id=282. Accessed on April 16, 2009.

Center for Genetics and Society. 2006b. "Research Cloning Arguments Pro and Con." Available online. URL: http://www.genetics andsociety.org/article.php?id=284. Accessed on April 17, 2009.

Center for Genetics and Society. 2009. "Failure to Pass Federal Cloning legislation, 1997–2003." Available online. URL: http://www.genetics andsociety.org/article.php?id=305. Accessed on April 17, 2009.

Cohen, Jon. 2009. "Stem Cells—Back by Popular Demand." Available online. URL: http://voices.washingtonpost.com/behind-the-numbers/2009/03/stem_cells_-_back_by_popular_d.html. Accessed on April 16, 2009.

Committee on Science, Engineering, and Public Policy. Board on Life Sciences. 2002. *Scientific and Medical Aspects of Human Reproductive Cloning*. Washington, D.C.: National Academies Press.

Connor, Steve. 2008. " 'Now we have the technology that can make a cloned child' " *The Independent*. April 14, 2008. Available online. URL: http://www.independent.co.uk/news/science/now-we-have-the-technology-that-can-make-a-cloned-child-808625.html. Accessed on April 16, 2009.

"Cosmetic Plastic Surgery Research: Statistics and Trends for 2001–2006." 2009. Available online. URL: http://www.cosmeticplastic surgerystatistics.com/statistics.html#2007-GRAPHS. Accessed on April 15, 2009.

Darus, Noormah Mohd, and Sin Lian Thye. [n.d.] *Enzyme Replacement Therapy for Metabolic Disorders*. Putrajaya, Malaysia: Health Technology Assessment Unit. Medical Development Division. Ministry of Health.

Economic Research Service. U.S. Department of Agriculture. 2009. "Adoption of Genetically Engineered Crops in the U.S." URL: http://www.ers.usda.gov/Data/BiotechCrops/. Accessed on March 19, 2009.

Federal Bureau of Investigation. 2009. "CODIS–NDIS Statistics." Available online. URL: http://www.fbi.gov/hq/lab/codis/clickmap.htm. Accessed on April 11, 2009.

Fletcher, Joseph. 1974. *The Ethics of Genetic Control: Ending Reproductive Roulette.* New York: Anchor Books.

Fletcher, Meg. 2007. "A Question of Privacy." IndustryFocus. Available online. URL: http://www.businessinsurance.com/cgi-bin/industry Focus.pl?articleId=21346&issueDate=2007-03-19. Accessed on April 15, 2009.

Food and Agriculture Organization. United Nations. 2004. *The State of Food And Agriculture 2003–2004: Agricultural Biotechnology. Meeting the needs of the poor?* Rome: Food and Agriculture Organization.

Fox, Jeffrey. 1998. "Germline Gene Theory Contemplated." *Nature Biotechnology.* 16 (5; May): 407.

Friedlander, Blaine P., Jr. 1999. "Toxic Pollen from Widely Planted, Genetically Modified Corn Can Kill Monarch Butterflies, Cornell Study Shows." Cornell News, May 19. Available online. URL: http://www.news.cornell.edu/releases/May99/Butterflies.bpf.html. Accessed on April 12, 2009.

"The Future of Food." 2004. A film available for viewing at http://www.snagfilms.com/films/title/the_future_of_food/. Accessed on April 12, 2009.

Nordlee, J. A. et al. 1996. "Identification of a Brazil Nut Allergen in Transgenic Soybeans." *New England Journal of Medicine.* 334 (11; March 14): 688–692

Genetics Home Reference. 2009. "Familial Adenomatous Polyposis." Available online. URL: http://ghr.nlm.nih.gov/condition=familialade nomatouspolyposis. Accessed on April 14, 2009.

Gillon, Raanan. 1999. "Human Reproductive Cloning—A Look at the Arguments Against It and a Rejection of Most of Them." *Journal of the Royal Society of Medicine.* 92 (1; January): 3–12.

Harris, J., and S. Chan. 2008. "Enhancement Is Good for You!: Understanding the Ethics of Genetic Enhancement." 15 (1; January): 338–339.

Institute for Responsible Technology. 2007. "Genetically Engineered Foods May Cause Rising Food Allergies." Available online. Part One: URL: http://www.seedsofdeception.com/utility/showArticle/?objectID=1007; Part Two: http://www.seedsofdeception.com/utility/showArticle/?objectID=1264. Accessed on April 11, 2009.

International Dairy Foods Association. 2007. "Basic Message Points on rbST." Available online. URL: http://www.idfa.org/reg/biotech/talking2.cfm. Accessed on April 12, 2009.

IU [Indiana University] News Room. 2003. "Plant Genes Imported from Unrelated Species More Often than Previously Thought, IU Biologists Find." Available online. URL: http://newsinfo.iu.edu/news/page/normal/1028.html. Accessed on April 11, 2009.

Keeling, Patrick J., and Jeffrey D. Palmer. 2008. "Horizontal Gene Transfer in Eukaryotic Evolution." *Nature Reviews Genetics*. 9 (8; August): 605–618.

Keim, Brandon. 2008. "Genetic Protections Skimp on Privacy, Says Gene Tester." Wired Science. Available online. URL: http://blog.wired.com/wiredscience/2008/05/genetic-protect.html. Accessed on April 14, 2009.

"Kennedy in Support of Genetic Information Nondiscrimination Bill." 2008. Press Release. Available online. URL: http://kennedy.senate.gov/newsroom/press_release.cfm?id=4FCF8E86-4706-4E74-B451-36253C5A425D. Accessed on April 15, 2009.

King, David. 1997. "Social Issues Raised by Gene Therapy." Available online. URL: http://www.hgalert.org/topics/hge/geneTherapy.htm. Accessed on April 15, 2009.

Kirk, Leigh. 2008. "WalMart Bans rBGH Tainted Milk." Natural News.com. Available online. URL: http://www.naturalnews.com/022913.html. Accessed on April 12, 2009.

Lambert, Bruce. 1998. "Giuliani Backs DNA Testing Of Newborns for Identification." *New York Times*. December 17.

Lander, Eric S., and Bruce Budowle. 1994. "Commentary: DNA Fingerprinting Dispute Laid to Rest." *Nature*. 371 (6500; October 27): 735–738.

Langridge, William H. R. 2000. "Edible Vaccines." *Scientific American*. 283 (3; September): 66–71.

Learn.Genetics. 2008. "Pharming for Farmaceuticals." Available online. URL: http://learn.genetics.utah.edu/archive/pharming/index.html. Accessed on April 12, 2009.

Lieberman, Adam J., and Simona C. Kwon. 1998. *Facts Versus Fears: A Review of the Greatest Unfounded Health Scares of Recent Times*, 3rd ed. Washington, D.C.: American Council on Science and Health. Available online. URL: http://www.solvaymartorell.com/static/wma/pdf/4/7/4/3/facts.pdf. Accessed on April 12, 2009.

Lovett, Kenneth. 2007. "Gov Bid to Win DNA Con-Test." *New York Post*. May 14, 2007. Available online. URL: http://www.nypost.com/seven/05142007/news/regionalnews/gov_bid_to_win_dna_con_test_regional news_kenneth_lovett___post_correspondent.htm. Accessed on April 11, 2009.

Making Sense of Your Genes: A Guide to Genetic Counseling. 2008. Chicago: National Society of Genetic Counselors, Inc.; Washington, D.C.: Genetic Alliance.

McGeehan, Patrick. 2007. "Spitzer Wants DNA Sampling In Most Crimes." *New York Times.* May 14, 2007. Available online. URL: http://query.nytimes.com/gst/fullpage.html?res=9905EFD91231 F937A25756C0A9619C8B63&sec=&spon=&pagewanted=all. Accessed on April 11, 2009.

McHughen, Alan. 2008. "Labeling Genetically Modified (GM) Foods." Available online. URL: http://agribiotech.info/details/McHugen-Labeling%20sent%20to%20web%2002.pdf.

Meade, Harry, and Carol Ziomek. 1998. "Urine as a Substitute for Milk?"*Nature Biotechnology.* 16 (1; January): 21–22.

Mellman Group. 2006. "Memorandum." November 16, 2006. Available online. URL: http://www.pewtrusts.org/uploadedFiles/ wwwpewtrustsorg/Public_Opinion/Food_and_Biotechnology/2006 summary.pdf. Accessed on April 12, 2009.

Moore, Solomon. 2009. "F.B.I. and States Vastly Expanding Databases of DNA." *New York Times.* April 19, 2009: 1, 19.

National Conference of State Legislatures. 2008. "State Genetic Privacy Laws." Available online. URL: http://www.ncsl.org/programs/ health/genetics/prt.htm. Accessed on April 14, 2009.

National Conference of State Legislatures. 2009. "State Human Cloning Laws." Available online. URL: http://www.ncsl.org/programs/ health/Genetics/rt-shcl.htm. Accessed on April 17, 2009.

National Newborn Screening and Genetic Resource Center. 2009. "National Newborn Screening Status Report; Updated 03/11/09." Available online. URL: http://genes-r-us.uthscsa.edu/nbsdisorders.pdf . Accessed on April 14, 2009.

Office of Science and Technology Policy. 1986. *Coordinated Framework for Regulation of Biotechnology.* 51 FR 23302, June 26, 1986. Available online. URL: http://usbiotechreg.nbii.gov/Coordinated_Frame work_1986_Federal_Register.html. Accessed on April 14, 2009.

Pearson, Yvette. 2006. "Never Let Me Clone?: Countering an Ethical Argument Against the Reproductive Cloning of Humans." *Science and Society.* 7 (7; July): 657–660.

PETA Medical Center. 2009. "Xenografts: Frankenstein Science." Available online. URL: http://www.peta.org/MC/factsheet_display.asp? ID=82. Accessed on April 14, 2009.

Pew Initiative on Food and Biotechnology. U.S. Food and Drug Administration. Cooperative State Research, Education and Extension Service, U.S. Department of Agriculture. 2002. *Pharming the Field: A Look at the Benefits and Risks of Bioengineering Plants to Produce Pharmaceuticals.* [n.p.]

Ramjoué, Celina. 2007. "The Transatlantic Rift in Genetically Modified Food Policy" Paper presented at the annual meeting of the International Studies Association 48th Annual Convention, Chicago, IL, February 28, 2007. Available online. URL: http://www.allacademic.com/meta/p179400_index.html

Religious Tolerance. 2007. "Therapeutic Cloning: U.S. Legislation at the Federal & State Level." Available online. URL: http://www.religioustolerance.org/clo_ther2.htm. Accessed on April 17, 2009.

"Responsa in a Moment: Genetic Engineering." 2009. Available online. URL: http://www.responsafortoday.com/moment/3_1.htm. Accessed on April 15, 2009.

Robinson, Clair. 2009. "Genetically Modified Justice?" Available online. URL: http://www.cropchoice.com/leadstry562e.html?recid=2135. Accessed on April 9, 2009.

San Diego Center for Molecular Agriculture. [n.d.]. *Foods from Genetically Modified Crops.* Available online. URL: http://www.sdcma.org/docs/GMFoodsBrochure.pdf. Accessed on April 12, 2009.

Seeds of Deception. 2009. "The Health Risks of GM Foods: Summary and Debate." Available online. URL: http://www.seedsofdeception.com/Public/GeneticRoulette/HealthRisksofGMFoodsSummaryDebate/index.cfm. Accessed on April 11, 2009.

State of New Mexico. 2009. "25-10-4. Genetic research program initiated." Available online. URL: http://www.conwaygreene.com/nmsu/lpext.dll?f=templates&fn=main-hit-h.htm&2.0. Accessed on April 17, 2009.

Stem Cell Information. 2009. "Stem Cells and Diseases." Available online. URL: http://stemcells.nih.gov/info/health.asp. Accessed on April 16, 2009.

Strong, Carson. 2005. "Reproductive Cloning Combined with Genetic Modification." *Journal of Medical Ethics.* 31 (11; November): 654–658.

Szebik, Imre, and Kathleen Cranley Glass. 2001. "Ethical Issues of Human Germ-cell Therapy: A Preparation for Public Discussion." *Academic Medicine.* 76 (1; January): 32–38.

The Campaign. 2009. Available online. URL: http://www.thecampaign.org/. Accessed on April 14, 2009.

Thompson, Larry. 2000. "Human Gene Therapy: Harsh Lessons, High Hopes." *FDA Consumer.* 34 (5; September—October): 19–24.

Timmerman, Luke. 2009. "Sticking to its Guns with Gene Therapy, Genzyme To Present Key Findings Within Days." Xconomy. Available online. URL: http://www.xconomy.com/boston/2009/03/25/sticking-to-its-guns-with-gene-therapy-genzyme-to-present-key-findings-within-days/. Accessed on April 15, 2009.

United Network for Organ Sharing. 2009. "It's All about Life." Available online. URL: http://www.unos.org/. Accessed on April 14, 2009.

U.S. Department of Health and Human Services. Food and Drug Administration. 1993. "Application of Current Statutory Authorities to Human Somatic Cell Therapy Products and Gene Therapy Products." *Federal Register.* 58 (197; October 14): 53248–53251.

U.S. Department of Health and Human Services. 1994. "Interim Guidance on the Voluntary Labeling of Milk and Milk Products From Cows That Have Not Been Treated With Recombinant Bovine Somatotropin." *Federal Register.* February 10: 6279.

U.S. Department of Health and Human Services. Food and Drug Administration. Center for Biologics Evaluation and Research. 1999. *Guidance For Industry. Public Health Issues Posed by the Use of Nonhuman Primate Xenografts in Humans.* Washington, D.C.: Food and Drug Administration.

U.S. Department of Justice. Federal Bureau of Investigation. 2009. "CODIS: Combined DNA Index System" [brochure]. Washington, D.C. : Federal Bureau of Investigation.

U.S. Food and Drug Administration. 2009. "Cellular & Gene Therapy Publications." Available online. URL: http://www.fda.gov/BiologicsBloodVaccines/GuidanceComplianceRegulatoryInformation/Guidances/CellularandGeneTherapy/default.htm. Accessed on April 15, 2009.

U.S. Government Accountability Office. 2006. *Nutrigenetic Testing: Tests Purchased from Four Web Sites Mislead Consumers.* GAO-06-977T.

"VCU Life Sciences Survey." 2008. Available online. URL: http://www.news.vcu.edu/doc/VCU-Life-Sciences-Report-2008.pdf. Accessed on April 16, 2009.

Willing, Richard. 2003. "White House Seeks to Expand DNA Database." *USA Today.* April 15, 2003. Available online. URL: http://www.usatoday.com/news/washington/2003-04-15-dna-usat_x.htm. Accessed on April 11, 2009.

Witten, Celia M. 2007. "Human Gene Therapies: Novel Product Development." FDA Consumer Update. Available online. URL: http://www
.fda.gov/consumer/updates/genetherapy101507.html. Accessed on
April 15, 2009.

Yu, Natalie Yu. "Edible Vaccines." 2008. *MMG 445 Basic Biotechnology
eJournal*. 4 (2008): 104–109.

3

Worldwide Perspective

Introduction

The United States is not Sweden. Or the United Kingdom. Or South Africa. Or New Zealand. New DNA technologies that have been developed—such as the genetic modification of food crops and animals, the development of genetic testing, and the introduction of human gene therapy procedures—has not affected all countries in the same way. Social and ethical issues that have the potential to tear apart the social fabric of one nation sometimes seem to have almost no effect on other nations. Technologies that have been subjected to strong regulations in some countries are left largely unfettered in others. This chapter offers a review of the way nations around the world other than the United States have responded to the availability of a host of new DNA technologies.

Forensic Science

The experience of many nations around the world with regard to DNA fingerprinting has been the same as that in the United States. Law enforcement officers have come to hold DNA testing in high regard, and the technology is widely acknowledged as being the "gold standard" in identification tools. As has also been the case in the United States, however, nations are struggling with the issue of DNA databases, attempting to decide how best to build such databases and how to use them in their own countries and in cooperation with other nations around the world.

Interpol, the international law enforcement agency, announced in 2007 that 42 member nations had some type of DNA database in addition to its own international database, known as DNA Gateway. Another 11 nations were in the process of developing DNA databases, and 30 more countries made use of DNA Gateway. At that time, the Interpol database contained 69,000 records submitted by 45 member nations. The system had produced to that point 145 international identifications in 13 different countries (Interpol 2007). European countries appear to have been most active in developing DNA databases. Efforts to coordinate law enforcement activity among the members of the European Union (EU) date back to 1988 with the creation of the European DNA Profiling Group. Efforts have continued with the formation of the Standardization of DNA Profiling in the European Union group in 2001 and a program of the European Network of Forensic Science Institutes in 2003 (Johnson and Williams 2004, 71).

The first concrete effort at meshing the efforts of two or more European nations was formalized in the Prüm Treaty, named after the German city where it was negotiated and signed on May 27, 2005. Signatories of the treaty—Austria, Belgium, France, Germany, Luxembourg, The Netherlands, and Spain— agreed to share their law enforcement databases, which included DNA specimens, fingerprint records, and vehicle registration data. The first trial run of this agreement involved an exchange of DNA information between Germany and The Netherlands in June 2008. As a result of that trial run, almost 600 hits in the Dutch database were obtained for German law enforcement officers, and more than 1,000 hits in the German database for their Dutch counterparts (Töpfer 2008, 14). The precise value of the system was not entirely verified by this experiment, since the vast majority of hits involved property crimes or anonymous "stains" from unidentified individuals left at crime scenes. Nonetheless, the Prüm signatories decided to go ahead with the implementation of the treaty provisions, and 10 additional countries—Bulgaria, Finland, Greece, Hungary, Italy, Portugal, Romania, Slovenia, Slovakia, and Sweden—decided to join the DNA consortium, which has since been transferred to the European Union's formal legal framework (Töpfer 2008, 14).

One of the fundamental problems with the joint DNA database arrangement in Europe has been, as it has been in the United States, a decision as to which individuals should be included. Standards have varied widely across nations and have included any recordable offense (England); any crime with a potential sentence of more than eight years (The Netherlands) or more than two years (Sweden); any serious crime upon conviction (Belgium, Norway); certain crimes specified in DNA enabling legislation (France, Germany); any suspected recordable offense (Austria, Slovenia, England); any arrest for a serious crime or sexual crime punishable by more than a year in prison (Finland, Germany), by more than 1.5 years in prison (Denmark), by more than five years in prison (Hungary), or in association with a history of serious offenses (Switzerland). Nations also differ significantly on the issue of permission to remove DNA samples from a database. Some nations permit removal of a convicted offender's DNA specimen after a period of 5 to 20 years (Belgium, Denmark, Germany, Hungary, The Netherlands, Switzerland, Sweden, Slovenia), while other nations retain those specimens forever (Austria, England, Finland, Norway). Some nations remove DNA specimens from a database if a suspect is not convicted of a crime or never charged with a crime (Austria, Denmark, Finland, France, Germany, Hungary, Finland, and Switzerland), while England retains those specimens forever. (Asplen 2003).

The diversity of these national systems is something of an issue for civil rights advocates since it means that someone required to provide a DNA specimen in one country with higher or lower standards might then be identified as a suspect in another country with very different standards. The problem is especially relevant to citizens of England who, even when innocent of any crime, may still have their DNA samples stored in England's national DNA database. Once England joins other European countries in a joint database, then, those innocent individuals may have their DNA information freely available to all other nations in the international database (Mail Online 2007).

These problems appear not to have dampened law enforcement officers in their desires to integrate DNA databases worldwide. In January 2007, the German Minister of the Interior Wolfgang Schäuble indicated that he was willing to begin discussing the possibility of sharing Germany's DNA database with the United States (Smith 2007). By July of the same year, Interpol announced that it had conducted its first successful test of DNA information sharing, using its sophisticated I-24/7 global law enforcement communications system. With that system it should soon be possible to transmit DNA information on any individual in any national database system to any other system in the world (Cooney 2007).

Genetically Modified Organisms

As in the United States, concerns about the testing, cultivation, transport, and sale of genetically modified foods in Europe date back to the 1980s, when a relatively small number of individuals began to foresee the possible commercial applications of DNA technology in the development of transgenic plants and animals in agriculture. At first, these concerns were limited to this small group of knowledgeable and politically active individuals and organizations. In response to these concerns, the Organisation for Economic Co-operation and Development (OCED) initiated a study on safety guidelines for the use of genetically modified organisms in agriculture and industry in 1983. Three years later, OCED published a set of guidelines based on this study (Organisation for Economic Co-operation and Development 1986.). Although the guidelines were nonbinding, European nations found them useful in developing their own individual policies.

The issue of genetically modified foods took on a greater relevance only two years after publication of the OCED guidelines. In 1988, four U.S. biotechnology companies applied for a European license to market genetically engineered bovine somatotropin hormone (rBSH) to increase milk production in cows. Worries about the introduction of GM foods spread rapidly across the continent, and a number of groups argued against approval of the rBSH request or any other request to introduce GM foods to the continent. In response to these concerns, the European Commission, the executive branch of the European Union, issued two directives in 1990 dealing with genetically modified organisms. The first directive, 90/219, dealt with the contained use of genetically modified microorganisms, while the second, 90/220, dealt with the intentional release (as in agricultural projects) of genetically modified organisms (Directive 90/219/EEC and Directive 90/220/EEC 1990).

The two directives were largely unsuccessful as an effort to standardize and unify the approval process for genetically modified foods. While some countries attempted to follow the guidelines closely, other nations tended to ignore recommendations and develop their own systems for regulation and approval. Between 1994 and 1998, the European Union itself had approved nine genetically modified crops, primarily varieties of corn, soybeans, and oilseed crops (United States Mission to the European Union 2003). But individual nations had begun to express their own concerns about the future of GM foods. In 1997, for example, Austria banned a genetically modified variety of corn that had already been approved by the European Union. When the Union took no action on Austria's decision, other countries followed Austria's lead. In 1997, Luxembourg also banned an EU-approved corn, followed in 1998 by a French ban on two varieties of rapeseed and a Greek ban on GM rapeseed, in 1999 by an Italian ban on four varieties of corn, and in 2000 by a German ban on GM corn. In all cases, the banned products had already been approved by the EU itself (United States Mission to the European Union 2003).

By June 1998, environmental ministers of the EU had adopted an informal, *de facto* ban on all GM foods. That ban applied both to the planting of genetically modified crops within the European states and to the import of GM foods to Europe. Those nations and individuals who support this ban have argued their case on the basis of the *precautionary principle*, which says that governmental agencies may be justified in taking certain regulatory actions even when some scientific uncertainty remains about the possible risks and consequences of a practice. In other words, "better safe than sorry" if the consequences of a practice are not well known and generally agreed upon. Over the next few years, a number of efforts were made by the EU itself to invalidate the ban on GM foods in some member countries and to allow both planting and importation of such products, but all such efforts have thus far failed (Harrison 2009).

The inability of the European Union to remove bans on GM foods on the continent became an increasing source of irritation to the United States government, which argued that this inaction was a violation of free-trade agreements between the United States and the European Union. Those agreements, U.S. representatives claimed, prohibited bans by any nation on the free flow of products among signatories to the agreements. In early

2003, the administration of President George W. Bush filed a formal complaint about EU practices with the World Trade Organization (WTO).

By the time the Bush administration took this action, however, the EU had already been forced into rethinking its position on GM foods. In March 2000, the European Court of Justice, the EU's highest judicial body, ruled that France had improperly banned three GM foods that had previously been approved by the EU (Judgment of the Court 21-03-2000 2000). Gradually, member states began to adopt the position that GM foods could be grown and imported *provided* that they met certain labeling and traceability standards. These standards were ultimately expressed in directive 2001/18 of the European Community, whose two main provisions are that (1) foods containing more than 0.9 percent genetically modified organisms must be so labeled, and (2) all GM foods must contain a *traceability tag*, a piece of DNA that contains the "address" of the company by whom it was produced (Directive 2001/18/ec of the European Parliament and of the Council of 12 March 2001 2001). A traceability tag allows a governmental agency to track down the manufacturer of a product that is found to have some deleterious effect on human health or the environment.

In May 2004, the ban on GM foods in the European Union officially came to an end with approval by the European Commission of a GM corn made by the Swiss company Syngenta (Meller and Pollack 2004, C1). Two years later, the EU set standards for the acceptance of GM foods that included two requirements. First, each product submitted for approval had to be shown to be safe for humans, other animals, and the environment. Second, every product had to be available on a "freedom of choice" basis. That is, every farmer and consumer was to have complete freedom in deciding whether or not to use a GM crop or food. Manufacturers were required to provide data needed to allow people to make those decisions (Environment News Service 2006).

Today, genetically modified foods and feed are readily available in the European Union, although there are still vigorous efforts underway to keep the continent "GMO-free." As something of an afterthought, the World Trade Organization finally ruled on the U.S. complaint of 2003 on May 16, 2006. The organization decided in favor of the United States and its co-complainants Argentina, Australia, Brazil, Canada, India,

Mexico, and New Zealand, indicating that the European Union's long ban on GM foods and feed was an infringement of free-trade agreements among all parties. The WTO also made a special point of criticizing the five European nations—Austria, France, Germany, Luxembourg, and Greece—that still maintained bans on the importation or growing of GM foods and feeds, in spite of contrary actions by the European Union (EurActive 2006).

The status of genetically modified crops in parts of the world other than the United States and Europe is mixed. On the one hand, the availability of GM foods and feed would appear to be a very important factor in efforts to feed the world's hungry. And, in fact, agricultural leaders in some of the poorest countries of the world make a plea for developed nations to cultivate a more positive attitude toward GM technology and to make that technology available to third world nations (Vallely 2009). But the acceptance or rejection of GM foods and feed appears to have almost nothing to do with food shortages or abundances in a nation. In 2003, for example, Zambia faced a national crisis when drought destroyed most of the grain crops on which its population depends for survival. But the Zambian government refused to accept shipments of grain from other nations that may have included genetically modified crops, choosing instead to let its people suffer from hunger and starvation. The government explained that it was not convinced that genetically modified foods were safe for humans and animals to consume (Manda 2003, 5).

In fact, only three nations in Africa currently permit the growing of genetically modified crops: South Africa, Burkina Faso, and Egypt. The latter two nations planted their first GM crops in 2008, each less than 50,000 hectares in size. In Asia and the Middle East, only China and India have substantial areas of land devoted to GM crops, 3.8 million hectares and 7.6 million hectares respectively (International Service for the Acquisition of Agri-biotech Applications 2008, Figure 1). Latin America represents a mixed situation for genetically modified crops, with slow expansion of the amount of land devoted to such crops and a growing and vigorous opposition developing to this trend (Ruiz-Marrero 2007). Argentina (21.0 million hectares) and Brazil (15.8 million hectares) have the second and third largest commitments to GM crops after the United States, but of the other countries only Paraguay (2.7 million hectares) has more than a million hectares of land under cultivation for GM crops. And in Central America, Honduras is the only country that grows GM crops, although farmers there appear to embrace the technology enthusiastically (Charles 2009; International Service for the Acquisition of Agri-biotech Applications 2008, Figure 1).

Public opinion about the safety and desirability of genetically modified foods is, understandably, an issue of importance to both proponents and opponents of the technology. Dozens of polls have been conducted in the United States, European nations, and other countries around the world. Not surprisingly, the results of those polls appear to differ significantly from nation to nation and even from year to year within any one nation. A 2000 public opinion poll in Japan found, for example, that GM crops received widespread support from respondents, ranging from about 40 percent for engineered cows that produce more milk to more than 50 percent for disease-resistant crops. Yet, these results showed some slackening in support over earlier polls on the same questions (Macer and Ng 2000, 945–947).

A study of particular interest was conducted by the International Service for the Acquisition of Agri-biotech Applications, the SEAMEO Regional Center for Graduate Study and Research in Agriculture, and the College of Development Communication, University of the Philippines Los Baños in 2006 (Torres, Suva, Carpio, and Dagli 2006). This study appears to be the most extensive and detailed study conducted anywhere on attitudes of businessmen and traders, consumers, extension workers, farm leaders and community leaders, journalists, policy makers, religious leaders, and scientists. The study assessed attitudes about GM foods and related topics in biotechnology and ascertained where the respondent groups get their information about biotechnology, the extent to which they trust various groups of decision-makers, their willingness to pay for engineered groups, and a host of other related issues. The introduction to the study also provides an excellent review of other public opinion surveys conducted in Asia.

One of the best indications of attitudes about GM foods in Europe is a series of polls conducted in 1991, 1993, 1996, 1999, 2002, and 2005 for Eurobarometer, the Public Opinion Analysis sector of the European Commission. These polls used samples of about 25,000 individuals, approximately 1,000 from each state in the European Union. The most recent of the polls (2005) found that 27 percent of respondents supported GM foods. This number masks somewhat the wide divergence of views across the continent, with the greatest support found in the Czech Republic (46 percent), Portugal (38 percent), and Finland (35 percent), with the lowest level of support in Luxembourg (13 percent), Greece (14 percent), Latvia (15 percent), and Cyprus (15 percent) (Gaskell, et al. 2006, 19).

Pollsters attempted to obtain more detail about the basis for people's attitudes about GM foods and crops. They identified three categories of individuals. First were those whom they called *outright supporters*, who thought that GM crops and foods are not risky, are morally acceptable, and potentially useful. A second group of respondents was identified as *risk tolerant supporters*, who viewed GM foods and crops as risky, but thought their use was morally acceptable and worthy of support. The third group of respondents was labeled simply *opponents*, who think that GM products are risky, morally unacceptable, and not potentially useful. In the 2005 survey, 58 percent of respondents fell into the "opponents" category, 25 percent into the "outright supporter" category, and 17 percent into the "risk tolerant supporter" category. The interesting comparison here is with the three other kinds of biotechnology included in this poll, nanotechnology, gene therapy, and pharmacogenetics, for whom the "opponents" categories included 9, 20, and 10 percent of respondents, respectively (Gaskell, et al. 2006, 20). Clearly, Europeans have much stronger attitudes about GM foods than about other types of biotechnology.

Opinions expressed in this 2005 poll tend to reflect those of other studies conducted in Europe (see, for example, Hoban 2004). But some recent polls suggest that people's *attitudes* about GM foods do not necessarily affect their *behavior* in shopping. A poll conducted for the European Commission between May 1, 2006 and October 30, 2008 found that nearly half of the people who bought GM foods said that they would not buy such foods, and another 30 percent said that they did not know whether they had been buying GM foods or not. And, perhaps in a somewhat discouraging finding for EU policy makers, a majority of shoppers are not able to distinguish GM foods from non-GM foods, primarily because they do not read the labels that policy makers so assiduously worked to develop as a way of resolving the debate over GM foods in the EU (Moses 2008, 17–7, 17–10). The irony of this finding is that, after more than a decade of intense debate and acrimony among government officials in member states of the European Union, consumers seem to actually care very little about the possible risks and morality of producing and selling GM foods; they have become just another group of items at the local supermarket, not all that different from the ones that have been on the shelves for years.

Pharming

In the European Union, the general principles by which plant-made pharmaceuticals (PMPs) are regulated are contained in two older directives, Directive 2001/18/EC, which deals with the deliberate release into the environment of genetically modified organisms, and Directive 2001/83/EC, which deals with medicinal products for human use. Both directives were published in 2001 and provide overall guidelines for GMOs and medical products, respectively, and apply to PMPs only by implication. In July 2008, the European Medicines Agency (EMA) released a new, specific directive addressing the regulation of PMPs specifically, "Guideline on the Quality of Biological Active Substances Produced by Stable Transgene Expression in Higher Plants." The guideline was more than two years in preparation, and came into effect on February 1, 2009. In its introduction to this document, the EMA noted that molecular farming is a very new technology, which suggests that those involved in research and development in this area should use special caution in their activities. The suggestions made in the guidelines are those that "should be employed in order to achieve satisfactory quality for biological active substances proposed to be produced using the new technology (European Medicines Agency 2008, 3/11).

The guidelines deal with four aspects of the production of PMPs: development, manufacturing, control of the active agent, and contamination with adventitious agents. Some of the topics discussed within each of these categories are the host plant, the transgene, transgenic banking systems, genetic stability, general manufacturing issues, characterization of the active agent, and control of possible adventitious agents.

An important milestone was reached in June 2006, when the European Medicines Agency approved the first transgenic drug for use in the European Union. That drug, Atryn®, had already been approved for use in the United States by the U.S. Food and Drug Administration. Atryn® is an anti-clotting agent and was approved for use in surgery for people with a rare congenital disease, antithrombin deficiency. It is obtained from goats who have been genetically modified to produce the chemical.

Nations other than those in the European Union have also developed regulatory systems for the control of research, development, and/or commercial production of PMPs. In Canada, for example, the earliest regulations on pharming were contained in a regulatory directive, "Conducting Confined Research Field Trials of Plant with Novel Traits in Canada," released in 2000. That directive dealt with issues such as the importation of plants with novel traits (PNTs), environmental safety assessments of PNTs for confined releases, risk management, and the production of data for environmental safety assessments of PNTs for unconfined releases (Canadian Food Inspection Agency 2000). This original directive has since been amended and updated a number of times (Canadian Food Inspection Agency 2006). As in the United States, responsibilities for the approval of PMPs is divided, with the Canadian Food Inspection Agency responsible for the plant crop itself and Health Canada responsible for the safety and efficacy of any products made from the plants. As of early 2009, no plant-made pharmaceutical had as yet been approved for use in Canada, nor had any genetically modified plant for the production of such pharmaceuticals yet been approved for commercial field use.

In Australia, as in most other countries, plant-made pharmaceuticals are regulated by a variety of governmental agencies that control various aspects of their research and development, testing, and commercialization. The primary overall document regulating pharming practices in the country is the Gene Technology Act of 2000, although each specific product and test is dealt with on a case-by-case basis by one of the various governmental agencies involved in the regulation of biotechnology (Bio-Regs Online 2006; Spök, et al. 2008). South Africa has a regulatory scheme for PMPs very similar to that of other developed nations with one important exception. The agencies responsible for approving genetically modified products in South Africa are empowered not only to assess the scientific and technological issues involved in the testing and development of PMPs (as is generally the case), but also to take into consideration social, economic, and other nonscientific issues that might impact communities in the vicinity of a field test or release of GMOs (Ma, et al. 2005).

A number of studies have been conducted inquiring into public opinion about the development and use of plant-made pharmaceuticals. One of the most recent and most complete of those studies was reported and discussed in a 2008 book, *Pharming: Promises and Risks of Biopharmaceuticals Derived from Genetically Modified Plants and Animals*. The survey reported here was based on 1,500 face-to-face interviews with residents of 12 members of the European Union (Austria, Czech Republic, Denmark, France, Germany, Ireland, Italy, The Netherlands, Poland, Spain, Sweden, and the United Kingdom), Israel, Japan, and the United States. Researchers first asked the extent to which respondents were familiar with the subject of PMPs. They found that, on average, between 20 and 35 percent of respondents felt that they were familiar with the issue. The highest levels of awareness were in Japan (33.6 percent) and Sweden (41.9 percent), and the lowest levels in Ireland (20.9 percent), The Netherlands (21.0 percent), and Italy (21.2 percent) (Rehbinder 2008, 153).

In the main body of their survey, researchers asked respondents to express their attitudes about a variety of statements about pharming, using a scale from 0 to 10, with 0 being total disagreement and 10 being total agreement. Some examples of these statements are: The genetic modification of plants in order to produce pharmaceutical drugs

- poses serious risks.
- is immoral.
- is very useful.
- is reckless.
- goes against nature.
- is the product of arrogance of scientists.
- is playing God.
- is the product of the interests of a small number of pharmaceutical multinationals.
- should be supported. (Rehbinder 2008, 154–155)

Probably the most interesting aspect of the responses to these questions was that some nations appear, in general, to be strongly opposed to the development of PMPs, while other nations are considerably more open to the process. Nations in the first category are Austria, Germany, and Japan, while those in the second category are Denmark, Israel, The Netherlands, and the United States. As an example, when asked if research on PMPs should be supported, respondents from Austria and Germany had the lowest approval scores, 4.2 and 4.6, respectively, while those in Israel and the United States had the highest scores, with a rating of 6.4 each (Rehbinder 2008, 155).

The final two questions dealt with somewhat different issues. In the first of those questions, researchers asked the extent to which respondents trusted laws and controls in their own nation on the regulation of pharming. The lowest scores on this issue came from respondents in Japan and Germany, whose scores were 4.4 and 4.8, respectively, while the greatest confidence in national controls were expressed by respondents in The Netherlands (6.1) and Denmark (6.8) (Rehbinder 2008, 156). Finally, researchers asked respondents if they, themselves, would be willing to take pharmaceuticals produced by engineered plants. Those most willing to do so were respondents from Denmark (7.0), The Netherlands (6.8), Czech Republic (6.5), and Israel (6.5), while the most wary respondents were those from Austria (3.8), Japan (4.0), France (4.8), Italy (4.8), and Germany (4.8) (Rehbinder 2008, 156). The results of this survey suggest that, while residents in some nations appear supportive of pharming technology, there is still significant and widespread doubt about the morality and desirability of this technology for the manufacture of pharmaceutical drugs.

Genetic Testing

Taking the pulse of Europeans on a variety of biotechnological issues has become something of a cottage industry on the continent, with surveys asking how people feel about genetically modified foods and crops, genetic testing, gene therapy, cloning, and other issues appearing regularly in the scholarly and popular press. In 2000, for example, George Gaskell, professor of social psychology at the London School of Economics and Political Science, asked more than 16,000 individuals in all nations belonging to the European Union their views on a number of issues in biotechnology. One of the interesting results is that, while people from different nations often had very different views about most technologies, genetic testing appeared to be by far the most widely accepted and approved of those technologies. In a chart summarizing each nation's views on seven technologies, genetic testing was viewed most positively in all 15 nations, with it ranking significantly more favorable in six of those nations (France, Greece, Italy, Netherlands, Poland, and Spain) (Gaskell 2000, Table 1). Another study conducted five years later produced similar results, with a general willingness across the European community to embrace the benefits that genetic testing has to offer. In the later survey, an additional finding was that nations that recognized specific benefits to be gained from genetic testing were more likely to favor the practice than those that saw fewer specific applications. In this survey, Belgium, France, Malta, and Portugal expressed the greatest support for genetic testing, while Austria, Germany, and Greece were least enthusiastic about the technology (Gaskell, et al. 2006, 53).

As is almost inevitably the case in Europe, various nations have a range of policies and practices about the use of genetic testing. In the case of one important issue, the testing of minors, for example, some nations have no laws or regulations at all on this point, other than those covered by other national health laws. Denmark and Germany are two such nations. Other countries have very strict laws that require specific and expressed public permission for testing to occur, often only after detailed information is provided by medical authorities. This situation holds in France, for example. The availability of genetic information obtained to outside sources, especially insurance companies, is also a matter of concern in many states. Some nations (Denmark and Spain, for example) have specific legislation prohibiting the provision of genetic test results to insurance companies. Most countries have no such laws, the United Kingdom, Germany, and Portugal among them (Bionet 2002).

In an effort to bring some cohesiveness to the European Union's position on genetic testing, a conference was held in Brussels on May 6 and 7, 2004, to discuss the status and future of genetic testing in the EU. A lengthy report of that conference included 25 recommendations for the use of genetic testing, categorized as "general framework," "implementation of genetic testing in healthcare systems," and "genetic testing as a research tool" (McNally and Cambon-Thomsen 2004a). Among those recommendations were the following:

- that genetic testing should be an integral part of conventional public health services; that it should never be imposed on an individual; and that information about genetic testing should be made readily available to citizens;
- that the European Union adopt a set of regulations to ensure that standards exist and are followed for all genetic testing;
- that individualized counseling be made available to anyone interested in the possibility of having genetic testing; and
- that data obtained from genetic testing never be used to discriminate against individuals or their families in matters of employment, insurance, and access to all social services.

Genetic testing policies and practices differ not only within the European Union, but also around the world. The Organisation for Economic Co-operation and Development (OECD) provides updates on the status of genetic engineering in many countries around the world on a regular basis. As an example, genetic testing technology is highly developed and widely used in Israel, with both prenatal and postnatal testing and genetic counseling available. There are currently no laws restricting the dissemination of information from these tests to outside agencies, such as insurance companies, although such a law is currently (2009) under consideration in the Knesset. In Australia, genetic testing services are available, but there is little formal governmental regulation of facilities or practices regarding testing and uses of test results.

A review of genetic testing throughout the world is beyond the scope of this book. Suffice it to say that many countries in which it might be of the greatest value are just those countries lacking in financial resources and medical facilities capable of employing genetic testing on a widespread and regular basis. As just one example, an estimated 200 million people around the world carry a gene for some type of pathologic hemoglobinopathy, which can result in a number of serious or fatal diseases, such as sickle-cell disease and thalassemia. The condition is easily detected with genetic tests, but such tests are generally not available to most people in Third World nations. Where genetic testing is available, it is sometimes used for less than desirable purposes, such as determining the sex of an unborn child. In countries where male babies are prized over female babies, the results of a genetic test may too often result in an abortion of the female fetus (Penchaszadeh 1993, 485–495).

Human Gene Therapy

In Europe, as in the United States, medical researchers began to think about the potential benefits of gene therapy long before technology was available to conduct clinical trials in the field. Those trials began in Europe in 1993 and have been continued ever since, with well over 600 clinical trials having been initiated and/or completed since that time (Clinigene-NoE 2009). Those trials have covered a wide variety of medical conditions, including amyotrophic lateral sclerosis, cystic fibrosis, glioblastoma, human immunodeficiency virus (HIV) disease, metastatic cancers, breast cancer, lung cancer, melanoma, neuroblastoma, adenosine deaminase deficiency-severe combined immunodeficiency disorder, colon adenocarcinoma, and brain tumors (Clinigene-NoE 2009).

Throughout this period of clinical testing, member nations and the European Union itself have struggled to develop regulatory frameworks for research on and clinical testing of gene therapy technologies. All nations have assigned responsibility for the regulation of gene therapy research and trials, even if no such activities in those areas have yet taken place. For example, in Iceland, the competent regulatory authorities for gene therapy are the Environment and Food Agency and the Icelandic Medicines Control Agency. In Malta, the competent authority is the Malta Environment and Planning Authority. In Slovenia, the relevant regulatory agency is the Agency for Medicinal Products and Medical Devices.

Nations in which research is well advanced and/or clinical trials have already been conducted generally have a more elaborate explication of policies and practices to be followed. In Germany, for example, gene therapy clinical trials are regulated by a number of laws, including the Embryo Protection Law, the Gene Technology Law, the German Medicines Act, the directive on the application of good clinical practice relating to the conduct of clinical trials on medicinal products for human use

(GCP directive), and certain provisions in the German Criminal Code (Gene Therapy Net.com 2009).

More than half of all gene therapy clinical trials currently take place in the United Kingdom, which began considering the technical and ethical implications of the technology in the late 1980s. In 1989, the Minister of Health appointed a special committee to consider the ethical implications of human gene therapy, chaired by Sir Cecil Clothier. The report of that committee, issued in 1992, has served as a general guideline for the conduct of research and clinical trials on gene therapy ever since (U. K. Department of Health 1992). Among the recommendations in the Clothier report was a suggestion that a supervisory body be established to oversee scientific and medical issues related to the safety and efficacy of gene therapy trials, and that this body work with local bioresearch ethics committees at institutions where trials are to be conducted. That body, the Gene Therapy Advisory Committee (GTAC) was established in 1993 and now serves as the primary regulatory agency for gene therapy research and trials in the United Kingdom. Researchers must submit detailed applications to the GTAC for all somatic cell gene therapy experiments; germ line cell experiments are prohibited in the United Kingdom (U. K. Department of Health 2009).

While individual nations were working to develop regulatory systems for human gene therapy, the European Union was also attempting to develop continent-wide guidelines and regulations. The earliest documents were a pair of directives dealing with the contained use of genetically modified organisms and the deliberate release of GMOs for medical and other purposes (Council Directive 98/81/EC 1998; Directive 2001/18/ec of the European Parliament and of the Council 2001). The European Parliament, the European Commission, and other agencies have since published a number of more detailed guidelines and regulations dealing with specific aspects of human gene therapy, including a consolidated code dealing with medicinal products for human use that provides detailed instructions for the use of genetically modified organisms in clinical trials, including human gene trial experiments (Directive 2001/83/ec of the European Parliament and of the Council 2001); a directive establishing specific provisions for the conduct of human gene therapy clinical trials (Directive 2001/20/ec of the European Parliament and of the Council 2001); a framework for the marketing of approved products and procedures used in human gene therapy (Regulation (Ec) No 726/2004 of the European Parliament and of the Council 2004); a general statement dealing with the risks of released GMOs into the germ line of organisms (General Principles to Address the Risk of Inadvertent Germline Integration of Gene Therapy Vectors 2006); and information on the use of viral vectors in gene therapy (Van der Akker, H.C.M. 2008; also see Gene Therapy Net.com 2009 for more detailed information on EU regulations on gene therapy).

Gene therapy research is not well advanced outside the United States and the European Union with one exception: China. In fact, the first gene therapy to receive government approval anywhere in the world was a treatment for head and neck squamous cell carcinoma called Gendicine, approved by the Chinese government in November 2003 (DrugResearcher .com. 2003). Since that time, patients from all over the world have come to China seeking treatment with the new technology, with somewhat mixed results. While the Chinese government is clearly enthusiastic about Gendicine and other gene therapy technologies, it provides significantly less regulatory oversight than do most other countries. This deficiency is a matter of some concern both in China and among researchers in other parts of the world (Business Week 2006). Nonetheless, China's very large patient population makes it a prime candidate for further advances in gene therapy, and as of 2006, the government had granted a total of 10 patents in the field, about a quarter the number issued in the United States (Tan 2005, e322). This progress has prompted some observers to call China "the world leader in gene therapy," although that accolade is probably true "simply because U.S. and European companies have much tougher safety and efficacy requirements" (Experimental Drugs Flourish in China 2008).

Gene therapy research in other parts of Asia, the Middle East, Africa, and Latin America has been slow. One important reason, of course, is the impoverished conditions and the lack of advanced medical research facilities in most nations. Nonetheless, some signs of progress are apparent. A group of researchers from the School of Medicine and University Hospital of the Universidad Autonoma de Nuevo Leon, in Monterrey, Mexico, the Advantagene Corporation, and the Methodist Hospital and Baylor College of Medicine announced in 2005 the first successful trial in Latin America of a gene therapy procedure for the treatment of prostate cancer. The treatment appeared to be successful, and other gene therapy procedures were in preparation (Rojas-Martinez, et al. 2005).

The best information currently available regarding public attitudes about gene therapy comes from a 2005 study by Eurobarometer, the public opinion arm of the European Commission. In this annual survey conducted in all member states, researchers asked a number of questions about the public's familiarity with and attitudes toward gene therapy. The survey's first finding was that Europeans, in general, are moderately familiar with the technology of gene therapy, with 45 percent of respondents indicating that they were "familiar with the technology." That number varied widely among member states, with citizens of The Netherlands (73 percent), Austria (68 percent), and Luxembourg (63 percent) apparently the best informed, while the least well informed were those in Malta (22 percent), Lithuania (32 percent), and the Czech Republic (32 percent) (Gaskell, et al. 2006, 16). The second theme of questions about gene therapy dealt with its moral acceptability, usefulness to society, potential for risk, and whether or not its development should be encouraged. Researchers found that gene therapy falls somewhere between nanotechnology and gene testing, both of which scored positively on all four factors, and genetically modified foods, considered the least morally acceptable, least useful to society, most risky, and least worthy of development. The conclusion they drew from these results is that Europeans tend to regard gene therapy as a "red" biotechnology, that is, a practice that involves not insignificant risks, but is probably worth those risks. Overall, 50 percent of respondents throughout the EU were supportive of gene therapy, with, again, the level of support ranging from highs of 60 percent (Belgium) and 58 percent (Cyprus and Italy) to lows of 32 percent (Malta and Lithuania) and 33 percent (Slovenia) (Gaskell, et al. 2006, 19).

Relatively few public opinion surveys about gene therapy or other topics in biotechnology have been conducted in the developing world. Some of those studies are now fairly old and of limited value. For example, the Eubios Ethics Institute, in Christchurch, New Zealand, conducted a comprehensive International Bioethics Survey in 1993, in which individuals in 10 countries—Australia, Hong Kong, India, Israel, Japan, New Zealand, The Philippines, Russia, Singapore, and Thailand—were asked about their attitudes toward a number of DNA technologies. That study found that respondents in China, India, and Thailand were significantly more optimistic about the promises of DNA technologies than were those in Europe and other countries in the survey. They were also less concerned about the practical and ethical issues raised by such technologies, such as the fear that scientists were "playing God" in their work (Macer, Azariah, and Srinives 2000, 313–332).

Cloning

As indicated in Chapter 2, opinions about and regulations of human reproductive cloning and therapeutic cloning differ dramatically. This situation is true in many parts of the world.

Reproductive Cloning

As in the United States, reproductive cloning is either discouraged or legally outlawed throughout the world. In many cases, nations rely on unwritten, but well-understood, cultural mores or on general, if sometimes somewhat vague, policy statements, laws, or regulations. For example, Mexico has no specific law dealing with reproductive cloning, but authorities have generally felt that the nation's Regulation on Scientific Research (RSR) and Regulation on the Sanitary Control of Organs, Tissues and Human Cadavers (RCOTHC) sections of Mexico's General Health Law (GHL) prohibit genetic manipulation of the germ line of an embryo (Genetics and Public Policy Center 2008). In many nations, relevant laws and regulations often refer to the need for protecting human life, health, and dignity in determining the types of research that may be conducted. The government of China, for example, promulgated a new policy statement on biomedical research in January 2007 for the purpose of "normalising the biomedical research involving human subjects and the application of technologies concerned to protect human life and health, safeguard human dignity, and protect legitimate rights and interests of human subjects" (Oxford Uehiro Centre for Practical Ethics 2007).

In a 2007 survey of 193 nations around the world, researchers for the Center for Genetics and Society found that 56 countries (29 percent) had passed specific legislation banning human reproductive cloning, while 137 nations (71 percent) had not yet adopted any such laws. The area in which such legislation was most likely to be found was Western Europe, where 17 nations (77 percent) had adopted laws banning human reproductive cloning, while only five—Andorra, Liechtenstein, Malta, Monaco, and San Marino—had not yet done so. In the Western Hemisphere, 10 nations (40 percent) have laws prohibiting reproductive cloning, while 25 (60 percent) do not. Africa and the Middle East have had by far the least interest in adopting formal legislation on reproductive cloning, with only South Africa, Tunisia, and Israel, of 76 nations in the area, currently having any formal legislation on the issue (Center for Genetics and Society 2007; for a comprehensive review of national legislation and regulations on reproductive cloning as of 2009, see Office for Human Research Protections. U.S. Department of Health and Human Services 2009; Javitt, Suthers, and Hudson 2006, 45–47).

Regional and international agencies have also taken action on reproductive cloning, making recommendations for legislation within a region or providing general standards for the behavior of member states on the issue. One of the most important of these documents is the *Additional Protocol to the Convention for the Protection of Human Rights and Dignity of the Human Being with regard to the Application of Biology and Medicine, on the Prohibition of Cloning Human Beings*, adopted by the Council of Europe in December 1998. The convention prohibits "any intervention seeking to create a human being genetically identical to another human being, whether living or dead." It also prohibits any exception to this ban, even in the case of a completely sterile couple (Council of Europe 1998, Article 1, Section 2). Framers of the document made clear that the purpose of the convention was to go beyond the specific issue of reproductive cloning and to "enshrine important principles which provide the ethical basis for further biological and medical developments, both now and in the future" (Council of Europe 2001). As of mid-2009, 17 nations had ratified the treaty and 14 nations had signed, but not ratified the treaty. It entered into force on January 3, 2001, when five nations had signed the treaty.

Perhaps the most fascinating statement on human reproductive cloning has been what has transpired with the United Nations. In 2000, France and Germany proposed that the United Nations convene a conference to write a convention that would ban human reproductive cloning, but that would leave therapeutic cloning for a later date. Objections to this proposal were raised by other nations, particularly Spain and the United States (as well as the Vatican City, acting as an observer). These countries lobbied for a more comprehensive treatment that would ban all kinds of cloning, both reproductive and therapeutic. Negotiations on the treaty dragged on for four years until a vote was held on March 8, 2005. The proposed draft of the treaty, banning all forms of cloning, was adopted by a vote of 84 in favor, 34 opposed, with 37 abstentions, and 33 nations absent from the vote. In the end, the vote had little significance not only because the final treaty consisted only of recommendations and was not binding on UN members, but also because the vote was so inconclusive (it passed with 45 percent of the total vote) (Center for Genetics and Society 2006).

In the first few years of the twenty-first century, a number of public opinion polls were conducted to test people's attitudes about reproductive cloning, with very similar results. When asked how they felt about reproductive cloning, whether they thought the government should ban reproductive cloning, whether reproductive cloning was morally acceptable or not, or some similar question, opposition to the practice was always very strong, usually ranging in the high 80–90 percent range. For example, when asked in 2000 whether they approved of or opposed scientists making genetic copies of humans, 90 percent of Canadians said they were opposed to the practice. A year later, when asked a similar question ("Are you for or against the cloning of human beings?"), 89 percent of Canadian respondents were again opposed to the practice. Two Gallup polls in 2001 of 13 nations, in one case, and 15 nations in the second found 81 percent and 93 percent of respondents in opposition to reproductive cloning. In only rare cases do polls show significantly less opposition to reproductive cloning. Two such polls (by the Discovery Channel in 2002, taken in Mexico and Turkey) found respondents about equally split on the issue, with about 50 percent in opposition to reproductive cloning (Center for Genetics and Society 2003).

Therapeutic Cloning

As indicated in the previous section, laws, regulations, and attitudes about therapeutic cloning tend to differ significantly from those about reproductive cloning. In general, regulations of cloning activities by a nation fall into one of three categories: (1) both forms of cloning are prohibited; (2) reproductive cloning is banned, but therapeutic cloning is permitted; and (3) reproductive cloning is banned, but no position is taken on therapeutic cloning, or any position taken is ambiguous. In one of the broadest surveys of international laws and regulations about cloning, researchers for the Genetics and Public Policy Center found in 2006 that all 45 countries surveyed had some type of restriction against human reproductive cloning, 28 of which also banned cloning for research purposes. Of the remaining 17 countries surveyed, 14 permitted therapeutic cloning, and three had indeterminate or ambiguous laws and/or regulations (Javitt, Suthers, and Hudson 2006, 45–47). The countries where therapeutic cloning is permitted are Belgium, China, Colombia, Hungary, India, Israel, Japan, Russia, Singapore, South Africa, South Korea, Sweden, Thailand, and the United Kingdom. An example of legislation that allows therapeutic cloning is that of Sweden, which first passed such legislation in 1991. At the time, the legislation was intended to deal with issues of in vitro fertilization, and it specified that embryos could be kept alive for no more than 14 days.

As the promise of embryonic stem cell research for therapeutic reasons became more clear, however, the Swedish government decided to push for an updated form of this legislation that would clarify and expand the conditions under which cloned embryos could be used for research purposes. That legislation, which was adopted in 2006, also carries specific restrictions on cloning for purposes of human reproduction (Ministry of Health and Social Affairs 2004). In India, efforts to develop a policy for therapeutic cloning research date back to the beginning of this century when the Department of Biotechnology of the Ministry of Science and Technology published a document, *Ethical Policies on the Human Genome*, that provided general guidance about the kinds of research that would and would not be permitted. Some observers regarded that document as incomplete and insufficiently clear, however, and the Indian Council of Medical Research began work on a more definitive statement on the subject almost immediately. In 2004, the council published a statement, *National Guidelines for Accreditation, Supervision and Regulation of ART Clinics in India, Ethical Issues and Consent Process Pertaining to Stem Cell Research* (2004), and *Draft Guidelines for Stem Cell Research and Therapy* (2006) (Genetics and Public Policy Center 2008). This combination of directives and guidelines is now regarded as one of the most complete and detailed regulatory statements about therapeutic cloning anywhere in the world (Isasi, et al. 2004, 626–640).

In general, Europeans tend to be supportive of therapeutic cloning, provided that it is regulated. In the 2005 Eurobarometer poll on a variety of opinions and attitudes of residents of the 25 member nations of the European Union, 52 percent said that they would approve the use of cloning for medical research purposes, 11 percent under all circumstances, and 41 percent provided that the procedure was highly regulated and controlled (Gaskell, et al. 2006, 81). As usual, opinions varied widely across nations, with the least support for therapeutic cloning coming from residents of Greece (never: 39 percent; only in exceptional circumstances: 21 percent), Austria (34 percent; 26 percent), and Finland (27 percent; 32 percent), and the greatest support coming from Sweden (65 percent support under all circumstances or when highly regulated and controlled), the United Kingdom (62 percent), and Belgium and Spain (59 percent) (Gaskell, et al. 2006, 85).

Public opinion polls about therapeutic cloning have also been conducted in a number of nations outside the European Union. Such a poll in Australia in 2006, for example, found that 58 percent of respondents supported therapeutic cloning (30 percent "strongly support" and 28 percent "somewhat support"), while 18 percent were totally opposed (10 percent "strongly opposed" and 8 percent "somewhat opposed") (Research Australia 2006, 13). A poll of Canadian public opinion in 2003 produced generally similar results, with 53 percent of respondents indicating that they supported therapeutic cloning overall and 32 percent opposing the practice. Perhaps the most interesting trend within this poll was the finding that a large majority of respondents between the age of 18 and 24 (66 percent) supported therapeutic cloning (Leger Marketing 2003, 3).

Conclusion

The potential applications of DNA technology are an issue of concern in many parts of the world. Developed nations in particular have been struggling with the best ways to take advantage of the benefits offered by these technologies, while limiting the risks they pose to human health and the environment. The

European Union, Australia and New Zealand, China, Japan, India, and South Africa have been especially involved in the writing of laws for the regulation and control of agricultural biotechnology, genetic testing, gene therapy, cloning, and related technologies. These laws tend to reflect the attitudes of citizens of the countries where they are enacted, ranging from moderately enthusiastic acceptance to significantly high levels of concern, depending on the specific technology. Somewhat unfortunately, many of the countries where these technologies have the greatest potential benefits—the developing nations of Latin America, Africa, the Middle East, and Asia—have taken few actions about the issues posed by DNA technology, at least partly because they do not have the medical, scientific, economic, political, and social resources with which to make use of the technologies. At some time in the future, these nations will probably inherit the promise and peril of DNA technologies, as the more developed nations in the world have already done.

References

Asplen, Christopher H. 2003. "International Perspectives on Forensic DNA Databases." A paper prepared for the ISRCL Conference, August 24–28, 2003, The Hague, the Netherlands. Available online. URL: http://www.isrcl.org/Papers/Asplenn.pdf. Accessed on April 17, 2009.

Bionet. 2002. "Who Owns Your Genes?" Available online. URL: http://www.bionetonline.org/english/content/gh_leg1.htm. Accessed on April 19, 2009.

BioRegs Online. 2006. *Australian Biotechnology Regulation Matrix*. Available online. URL: http://www.bioregs.gov.au/questions/images/overview_matrix.pdf. Accessed on April 28, 2009.

Business Week. 2006. "A Cancer Treatment You Can't Get Here." Available online. URL: http://www.businessweek.com/magazine/content/06_10/b3974104.htm. Accessed on April 20, 2009.

Canadian Food Inspection Agency. 2000. *Directive Dir2000-07: Conducting Confined Research Field Trials of Plant with Novel Traits in Canada*. Available online. URL: http://www.inspection.gc.ca/english/plaveg/bio/dir/dir0007e.shtml#a1_1. Accessed on April 28, 2009.

Canadian Food Inspection Agency. 2006. *Developing a Regulatory Framework for the Environmental Release of Plants with Novel Traits Intended for Commercial Plant Molecular Farming in Canada*. Available online. URL: http://www.inspection.gc.ca/english/plaveg/bio/mf/fracad/commere.shtml. Accessed on April 28, 2009.

Center for Genetics and Society. 2003. "CGS Summary of Public Opinion Polls." Available online. URL: http://www.geneticsandsociety.org/article.php?id=401#repro. Accessed on April 22, 2009.

Center for Genetics and Society. 2006. "The United Nations Human Cloning Treaty Debate, 2000–2005." Available online. URL: http://geneticsandsociety.org/article.php?id=338&&printsafe=1. Accessed on April 22, 2009.

Center for Genetics and Society. 2007. "National Polices on Human Genetic Modification: A Preliminary Survey." Available online. URL: http://www.geneticsandsociety.org/article.php?id=304. Accessed on April 22, 2009.

Charles, Dan. 2009. "Honduras Embraces Genetically Modified Crops." NPR. Available online. URL: http://www.npr.org/templates/story/story.php?storyId=93310225. Accessed on April 18, 2009.

Clinigene-NoE. 2009. "Clinigene European Clinical Trial Database." Available online. URL: http://141.39.175.7/eutrials/. Accessed on April 20, 2009.

Cooney, Michael. 2007. "New Messaging Network Exchanges/tracks DNA Data Worldwide." Layer 8. Available online. URL: http://www.networkworld.com/community/node/17511. Accessed on April 17, 2009.

"Council Directive 98/81/EC." 1998. *Official Journal of the European Communities.* December 5, 1998, L330/13-L330-31.

Council of Europe. 1998. *Additional Protocol to the Convention for the Protection of Human Rights and Dignity of the Human Being with regard to the Application of Biology and Medicine, on the Prohibition of Cloning Human Beings.* Available online. URL: http://conventions.coe.int/Treaty/en/Treaties/Html/168.htm. Accessed on April 22, 2009.

Council of Europe. 2001. *Additional Protocol to the Convention for the Protection of Human Rights and Dignity of the Human Being with regard to the Application of Biology and Medicine, on the Prohibition of Cloning Human Beings.* Available online. URL: http://conventions.coe.int/Treaty/en/Summaries/Html/168.htm. Accessed on April 22, 2009.

Directive 90/219/EEC and Directive 90/220/EEC. 1990. European Commission. Available online. URL: http://ec.europa.eu/enterprise/pharmaceuticals/eudralex/vol-1/dir_1990_219/dir_1990_219_en.pdf and http://www.biosafety.be/GB/Dir.Eur.GB/Del.Rel./90.220/TC.html. Accessed on April 18, 2009.

"Directive 2001/18/ec of the European Parliament and of the Council of 12 March 2001." 2001. *Official Journal of the European Communities*. April 17, 2001, L106/1-L106/38.

"Directive 2001/20/ec of the European Parliament and of the Council." 2001. *Official Journal of the European Communities*. May 1, 2001, L 121/34-L 121/44.

"Directive 2001/83/ec of the European Parliament and of the Council." 2001. *Official Journal of the European Communities*. November 28, 2001, L 311/67-L 311/128.

DrugResearcher.com. 2003. "First Gene Therapy Approved in China." Available online. URL: http://www.drugresearcher.com/Tools-and-techniques/First-gene-therapy-approved-in-China. Accessed on April 20, 2009.

Environment News Service. 2006. "EU Alters GMO Assessments to Satisfy Resistant Member States." Available online. URL: http://www.ens-newswire.com/ens/apr2006/2006-04-12-04.asp. Accessed on April 18, 2009.

EurActive. 2006. "EU GMO Ban Was Illegal, WTO Rules." Available online. URL: http://www.euractiv.com/en/trade/eu-gmo-ban-illegal-wto-rules/article-155197. Accessed on April 18, 2009.

European Medicines Agency. 2008. *Guideline on the Quality of Biological Active Substances Produced By Stable Transgene Expression in Higher Plants*. London, July 24, 2008. Available online. URL: http://www.emea.europa.eu/pdfs/human/bwp/4831606enfin.pdf. Accessed on April 28, 2009.

"Experimental Drugs Flourish in China." 2008. *China Business*. Available online. URL: http://www.atimes.com/atimes/China_Business/JA09Cb02.html. Accessed on April 21, 2009.

Gaskell, George. 2000. "Agricultural Biotechnology And Public Attitudes In The European Union." *AgBioForum*. 3 (2&3). Available online. URL: http://www.agbioforum.missouri.edu/v3n23/v3n23a04-gaskell.htm. Accessed on April 18, 2009.

Gaskell, George, et al. 2006. *Europeans and Biotechnology in 2005: Patterns and Trends*. Available online. URL: http://ec.europa.eu/public_opinion/archives/ebs/ebs_244b_en.pdf. Accessed on April 19, 2009.

Gene Therapy Net.com. 2009. "Gene Therapy Legislation in Germany." Available online. URL: http://www.genetherapynet.com/germany.html. Accessed on April 20, 2009.

"General Principles to Address the Risk of Inadvertent Germline Integration of Gene Therapy Vectors." 2006. Available online. URL: http://

www.emea.europa.eu/pdfs/human/ich/46999106en.pdf. Accessed on April 20, 2009.

Genetics and Public Policy Center. 2008. "International Law Search." Available online. URL: http://www.dnapolicy.org/policy .international.php?action=detail&laws_id=42. Accessed on April 22, 2009.

Harrison, Pete. 2009. "EU Upholds Austria, Hungary Right to Ban GM Crops." Available online. URL: http://www.reuters.com/article/ environmentNews/idUSTRE5212OL20090302. Accessed on April 18, 2009.

Hoban, Thomas J. 2004. *Public Attitudes towards Agricultural Biotechnology.* [Geneva]: Agricultural and Development Economics Division. The Food and Agriculture Organization of the United Nations.

International Service for the Acquisition of Agri-biotech Applications. 2008. "Global Status of Commercialized Biotech/GM Crops: 2008. The First Thirteen Years, 1996 to 2008." Available online. URL: http://www.isaaa.org/resources/publications/briefs/39/ executivesummary/default.html. Accessed on April 18, 2009.

Interpol. 2007. "INTERPOL Network Links Forensic Laboratories in G8 Countries. Available online. URL: http://www.interpol.com/Public/ ICPO/PressReleases/PR2007/PR200729.asp. Accessed on April 17, 2009.

Isasi, Rosario M., et al. 2004. "Legal and Ethical Approaches to Stem Cell and Cloning Research: A Comparative Analysis of Policies in Latin America, Asia, and Africa." *Journal of Law, Medicine, and Ethics.* 32 (4): 626–640.

Javitt, Gail H., Kristen Suthers, and Kathy Hudson. 2006. *Cloning: A Policy Analysis.* Washington, D.C.: Genetics and Public Policy Center.

Johnson, Paul, and Robin Williams. 2004. "DNA and Crime Investigation: Scotland and the 'UK National DNA Database'." *Scottish Journal of Criminal Justice Studies* 10: 71–84.

"Judgment of the Court 21-03-2000". 2000. Available online. URL: http://www.eel.nl/cases/HvJEG/699j0006.htm. Accessed on April 18, 2009.

Leger Marketing. 2003. *How Canadians Feel about Cloning.* Montréal: Leger Marketing.

Ma, Julian K-C, et al. 2005. "Molecular Farming for New Drugs and Vaccines: Current Perspectives on the Production of Pharmaceuticals in Transgenic Plants." *EMBO Reports* 6 (7; July): 593–599.

Macer, Darryl, and Mary Ann Chen Ng. 2000. "Changing Attitudes to Biotechnology in Japan: Support for Biotechnology in Japan Is

Declining, Although it Remains Higher than the US or Europe." *Nature Biotechnology.* 18 (9; September): 945–947.

Macer, D. R. J., J. Azariah, and P. Srinives. 2000. "Attitudes to Biotechnology in Asia." *International Journal of Biotechnology.* 2 (4): 313–332.

Mail Online. 2007. "Fears over innocent Britons' DNA being given to European police forces." Available online. URL: http://www.dailymail.co.uk/news/article-453739/Fears-innocent-Britons-DNA-given-European-police-forces.html. Accessed on April 17, 2009.

Manda, Olga. 2003. "Controversy Rages over 'GM' Food Aid." *Africa Recovery.* 16 (4; February): 5.

McNally, Eryl, and Anne Cambon-Thomsen. 2004a. *Ethical, Legal and Social Aspects Of Genetic Testing: Research, Development and Clinical Applications.* Brussels: European Commission. Available online. URL: http://ec.europa.eu/research/conferences/2004/genetic/pdf/report_en.pdf. Accessed on April 19, 2009.

McNally, Eryl, and Anne Cambon-Thomsen. 2004b. *25 Recommendations On the Ethical, Legal And Social Implications Of Genetic Testing.* Brussels: European Commission. Available online. URL: http://ec.europa.eu/research/conferences/2004/genetic/pdf/recommendations_en.pdf. Accessed on April 19, 2009.

Meller, Paul, and Andrew Pollack. 2004. "Europeans Appear Ready To Approve a Biotech Corn." *New York Times.* May 15, 2004: C1.

Ministry of Health and Social Affairs. 2004. *Government Bill 2003/04:148: Stem Cell Research.* Stockholm: Regeringskansliet. Available online. URL: http://www.sweden.gov.se/content/1/c6/02/63/38/387bbc1d.pdf. Accessed on April 22, 2009.

Moses, Vivian. 2008. *Do European Consumers Buy GM Foods?.* European Commission. Framework 6. Available online. URL: http://www.kcl.ac.uk/downloads/schools/biohealth/research/nutritional/consumer choice/full-part01.pdfand http://www.kcl.ac.uk/downloads/schools/biohealth/research/nutritional/consumerchoice/full-part02.pdf. Accessed on April 18, 2009.

Office for Human Research Protections. U.S. Department of Health and Human Services. 2009. *International Compilation of Human Research Protections 2009 Edition.* Available online. URL: http://www.hhs.gov/ohrp/international/HSPCompilation.pdf. Accessed on April 22, 2009.

Organisation for Economic Co-operation and Development. 1986. "Recombinant DNA Safety Considerations." Available online. URL: http://dbtbiosafety.nic.in/guideline/OACD/Recombinant_DNA_safety_considerations.pdf. Accessed on April 18, 2009.

Oxford Uehiro Centre for Practical Ethics. 2007. "Regulations and Laws." Available online. URL: http://www.chinaphs.org/bioethics/regulations_&_laws.htm#Top. Accessed on April 22, 2009.

Penchaszadeh, Victor B. 1993. "Reproductive Health and Genetic Testing in the Third World." *Clinical Obstetrics and Gynecology.* 36 (3; September): 485–495.

"Regulation (Ec) No 726/2004 of the European Parliament and of the Council." 2004. *Official Journal of the European Communities.* April 20, 2001, L 136/1-L 136/33.

Rehbinder, E., et al. 2008. *Pharming: Promises and Risks of Biopharmaceuticals Derived from Genetically Modified Plants and Animals.* Berlin: Springer Verlag.

Research Australia. 2006. *Health & Medical Research Public Opinion Poll 2006.* Sydney: Research Australia Limited.

Rojas-Martinez, Augusto, et al. 2005. "Gene Therapy Against Prostate Cancer Using the HSV-tk/Prodrug System Followed by Prostatectomy." *Molecular Therapy.* 11: 116–117.

Ruiz-Marrero, Carmelo. 2007. "Latin America: The Downside of the GM Revolution." *America's Program.* Available online. URL: http://americas.irc-online.org/am/4786. Accessed on April 18, 2009.

Smith, Robert W. 2007. "Interior Minister Proposes Pan-european Network of DNA and Fingerprint Databases." *Heise Online.* Available online. URL: http://www.heise.de/english/newsticker/news/83794. Accessed on April 17, 2009.

Spök, Armin, et al. 2008. "Evolution of a Regulatory Framework for Pharmaceuticals Derived from Genetically Modified Plants." *Trends in Biotechnology.* 26 (9; September): 506–517.

Tan, Chris Y. H. 2005. "Science Star over Asia." *PloS Biology.* 3 (9; September): e322.

Töpfer, Eric. 2008. "Searching for Needles in an Ever Expanding Haystack: Cross-border DNA Data Exchange in the Wake of the Prum Treaty." *Statewatch.* 18 (3; July-September): 14–16.

Torres, Cleofe S., Madeline M. Suva, Lynette B. Carpio, and Winifredo B. Dagli. 2006. *Public Understanding and Perception of and Attitude Towards Agricultural Biotechnology in the Philippines.* Los Baños, Philippines: International Service for the Acquisition of Agri-biotech Applications, SEAMEO Regional Center for Graduate Study and Research in Agriculture, and College of Development Communication, University of the Philippines Los Baños.

U. K. Department of Health. 1992. *Report of the Committee on the Ethics of Gene Therapy.* London Her Majesty's Stationery Office, 1992.

U. K. Department of Health. 2009. "Gene Therapy Advisory Committee (GTAC)." Available online. URL: http://www.advisorybodies .doh.gov.uk/genetics/gtac/. Accessed on April 20, 2009.

United States Mission to the European Union. 2003. "USTR Details Urgency of Ending EU Ban on Biotech Food." Available online. URL: http://useu.usmission.gov/Article.asp?ID=F43DE3D7-B6EE-4126-ABD1-557CAAC44727. Accessed on April 18, 2009.

Vallely, Paul. 2009. "Strange Fruit: Could Genetically Modified Foods Offer a Solution to the World's Food Crisis?" The Independent. Available online. URL: http://www.independent.co.uk/life-style/food-and-drink/features/strange-fruit-could-genetically-modified-foods-offer-a-solution-to-the-worlds-food-crisis-1668543.html. Accessed on April 18, 2009.

Van der Akker, H.C.M. 2008. *Environmental Risk Assessment of Replication Competent Viral Vectors in Gene Therapy Trials*. Bilthoven, the Netherlands: National Institute for Public Health and the Environment.

4

Chronology

This chapter provides a time line of events in the history of recombinant DNA research and its applications. It includes important scientific and technological developments, as well as social, political, and ethical issues related to DNA technology. The dates in this chapter are often approximate since, for example, a researcher may begin his or her studies in one year, continue them for many years, and report results at some later date or dates. Or, a law may be passed in one year, but not take effect until some time later.

Premodern times   Humans discover any number of basic principles in biotechnology for which they have no scientific explanation. These principles and processes arise as a result of trial-and-error efforts in the fields of agriculture, dairying, beer and wine production, fermentation, food preservation, and human growth patterns.

1859   Charles Darwin publishes *The Origin of Species*, one of the most influential books in biology ever written. The book provides an explanation for the transmission of characteristics from one generation to the next, although Darwin fails to provide a specific mechanism by which the process occurs.

1865   Gregor Mendel, an Austrian monk, presents his research on growth patterns in peas to the Natural Science Society of Brünn. He hypothesizes that characteristic traits are transmitted from one

1865 (*cont.*)	generation to the next by means of certain "factors" contained within plants. Mendel's so-called "factors" are now known as genes.
1868	Darwin hypothesizes that the unit of inheritance in plants and animals is a unit that he calls the gemmule. He is, at the time, unaware of Mendel's research or that his "gemmule" is similar both to Mendel's "factor" and to the modern-day gene.
1869	Swiss physician and biologist Johannes Friedrich Miescher discovers a previously unknown phosphorus-rich compound which he calls *nuclein*. The compound is later found to be a nucleic acid, although its biological function is not understood for many years.
1879	German physician and researcher Albrecht Kossel begins his research on the chemical composition of cellular components. He discovers that Miescher's nuclein is actually a class of biochemical compounds later given the name of *nucleic acids* by German pathologist Richard Altman in 1889.
1882	German biologist Walther Flemming announces his discovery of the process of mitosis, an event by which the genetic characteristics of two parents are transmitted to their offspring.
1883	English polyglot Francis Galton coins the term *eugenics* to improve the human race by selective breeding.
1884	Mendel dies without his accomplishments having been recognized in his lifetime.
1900	Three scientists, Dutch botanist Hugo De Vries, Austrian botanist Erich von Tschermak, and German botanist Karl Correns, independently rediscover Mendel's research in preparing for their own genetic studies.

1902	American geneticist and physician Walter Stanborough Sutton proposes the chromosomal theory of inheritance. He says that Mendel's "factors" are units located on chromosomes, which can explain his theory of independent assortment of genetic characteristics.
1903	Danish botanist and geneticist Wilhelm Johannsen suggests the germs *gene*, *genotype*, and *phenotype* to describe the unit of inheritance and the genetical and physical features of an organism produced by the transmission of genes.
1904	British geneticists William Bateson and Reginald Punnett discover the process of gene linkage, in which two genetic traits occur on the same chromosome and are, therefore, usually inherited together.
1906	Bateson suggests the word *genetics* to describe the science of inheritance.
1908	British physician and medical researcher Archibald Edward Garrod proposes the one gene/one enzyme theory, which says that each gene is responsible for the production of one specific protein (enzyme). Garrod's idea is largely ignored until it is rediscovered and proved by the research of American geneticists, George Beadle and Edward Tatum, in 1941.
1909	Russian-American biochemist Phoebus Levene discovers the presence of the sugar ribose in nucleic acids. Nucleic acids containing this compound later become known as *ribonucleic acids*, or RNAs.
1911	American geneticist and embryologist Thomas Hunt Morgan establishes the role of genes and chromosomes in the transmission of heritable characteristics in a series of ingenious experiments with fruit flies.

1913	American geneticist Alfred Sturtevant creates the first genetic map of a chromosome, one of the fruit fly's eight chromosomes.
1917	(Often misattributed to 1919) Hungarian engineer Karl Ereky coins the term *biotechnology* and uses the term to refer to any product made from raw materials with the aid of living organisms.
1922	American geneticist Hermann J. Muller publishes a remarkably prescient paper, "Variation Due to Change in the Individual Gene," in which he summarizes current knowledge about the chemical and physiological characteristics of the gene and its role in the inheritance of physical characteristics. At this point, however, scientists do not yet know what a gene is.
1929	Levene discovers a previously unknown component of nucleic acids, the sugar deoxyribose, in place of the sugar ribose already known to be a component of nucleic acids. He calls the new kind of nucleic acid *deoxyribose nucleic acid*, or DNA.
1934	British physicist John Desmond Bernal finds that X-ray crystallography can be a powerful tool in determining the molecular structure of proteins. The method later becomes an essential tool for finding the molecular structure of DNA.
1935	Russian biochemist Andrei Nikolaevitch Belozersky isolates DNA in the pure state for the first time.
1938	American engineer, mathematician, and science administrator Warren Weaver suggests the name *molecular biology* for the new field of science in which biological phenomena are studied and interpreted in chemical and physical (primarily molecular) terms.

1940	Belgian biochemist Jean Brachet hypothesizes that DNA and RNA in cells have two very different functions, and that the latter is actually involved in the production of protein molecules. Brachet's hypothesis presages the fundamental dogma of molecular biology, namely that DNA makes RNA, which makes proteins.
1941	American geneticists George Beadle and Edward Tatum rediscover and prove A. E. Garrod's one gene/one enzyme theory (1908).
	Danish microbiologist A. Justin (also given as Jost and Joost) is credited with first having used the term *genetic engineering* to describe the transfer of one piece of genetic material from one organism to a second organism.
1944	Medical researchers Oswald Avery, Colin MacLeod, and Maclyn McCarty hypothesize that DNA is responsible for transformations observed in succeeding generations of bacteria. They argue that DNA may play the role previously assigned to genes, in more complex organisms. The hypothesis is startling and largely rejected, because researchers are largely convinced that nucleic acids are too small and simple to carry genetic information (and that proteins are the more likely candidates for this role).
1946	German-American biophysicist Max Delbrück and American bacteriologist and geneticist Alfred Day Hershey independently find that the genetic material from two different viruses can combine with each other in such a way as to produce a third and new type of virus. Their experiments are an early example of the recombination of DNA that lies at the basis of much DNA technology today.
1950	Austrian-American biochemist Erwin Chargaff discovers that the quantity of nitrogen bases adenine and thymine is equal to that of the bases

1950 (*cont.*)	cytosine and guanine. Although there is no obvious significance to this discovery at the time, it provides a key piece of information needed by Watson and Crick in their elucidation of the structure of the DNA molecule (1953).
	British biophysicist Rosalind Franklin and New Zealand-born British molecular biologist Maurice Wilkins begin their X-ray crystallographic studies of DNA molecules and conclude that they exist as double-stranded helices, information that became a key to the solution of DNA structure by Watson and Crick in 1953.
1952	American molecular biologists Joshua Lederberg and Norton Zinder discover the process of bacterial transduction, in which a virus removes DNA from one bacterial cell and transports it to a second bacterium. The process, in principle, is one used by modern-day researchers to alter the genetic constitution of a cell, and it is still called by that name.
	Hershey and American geneticist Martha C. Chase (Epstein) carry out the famous "blender experiment," in which they conclusively demonstrate that DNA (and not protein) is the genetic material.
	American biologists Robert Briggs and Thomas King clone the first vertebrate by transplanting nuclei harvested from leopard frogs into eggs from which nuclei had been removed.
1953	American biologist James Watson and English chemist Francis Crick elucidate the structure of the DNA molecule, showing that it occurs as a double-stranded helical chain of sugar and phosphate groups linked by four nitrogen bases.
1955	American physicist, molecular biologist, and behavioral geneticist Seymour Benzer develops methods for studying the detailed molecular

characteristics of mutations and finds, among other things, that changes can take place at specific points within a DNA molecule, sometimes at the level of a single nucleotide.

1957   American geneticist and molecular biologist Matthew Meselson and American molecular biologist Franklin Stahl demonstrate the process by which DNA molecules self-replicate, a process described as *semiconservative*, because the two strands of the DNA molecule act as templates for the construction of two new identical DNA molecules without being destroyed.

1958   Francis Crick enunciates the Central Dogma principle of molecular biology, essentially that "once information has got[ten] into a protein it can't get out again." The Central Dogma further says that the most common type of information transfer in a cell is DNA $\rightarrow$ DNA $\rightarrow$ RNA $\rightarrow$ Protein, but under special circumstances, may include RNA $\rightarrow$ RNA, RNA $\rightarrow$ DNA, and DNA $\rightarrow$ Protein.

1959   French geneticist Jérôme Lejeune demonstrates that Down syndrome results from the presence of an extra copy of all or part of chromosome number 21, the first instance in which a genetic basis for a disease had been identified.

1960   Danish researchers Povl Riis and Fritz Fuchs develop a method for determining the sex of an unborn mammal by prenatal analysis of amniotic fluid.

1961   American microbiologist Robert Guthrie develops a method for screening the blood of a newborn child for the genetic disease phenylketonuria.

1966   Indian-American molecular biologist Har Gobind Khorana and American biochemists Marshall Nirenberg and Robert Holley elucidate the genetic

1966 (*cont.*)	code, the set of three nitrogen bases that code for the amino acids that make up proteins.
1968	Swiss microbiologist Werner Arber discovers restriction enzymes, compounds with the ability to recognize and cut DNA molecules at specific points.
1970	American molecular biologist David Baltimore and American geneticist Howard Temin independently discover the enzyme reverse transcriptase, which facilitates the transcription of RNA material into a new DNA molecule, a discovery with a host of applications in molecular biology, one of which is a method for producing clones.
	American microbiologist Hamilton O. Smith discovers the first site-specific restriction enzyme, a compound that recognizes a specific base sequence in a DNA molecule and then cuts the molecule at some point within that sequence.
1972	American biochemist Paul Berg develops a method for inserting a segment of DNA from a foreign molecule into a host DNA molecule, producing the first recombinant DNA molecule. The method is sometimes referred to as *gene splicing*. Berg and some colleagues soon write to the National Institutes of Health pointing out possible safety, social, and ethical issues related to this research, an act that eventually leads to the Asilomar Conference of 1975.
1973	American biochemists Stanley N. Cohen and Herbert Boyer develop a method for inserting foreign DNA segments into a plasmid and then incorporating the altered plasmid molecule into the DNA of a bacterium. The procedure results in the formation of an engineered organism (the bacterium) that can then make endless copies of itself (cloning), while displaying characteristics different from those of the original bacterium. The

experiment marks the beginning of the modern field of recombinant DNA procedures.

1974   The National Institutes of Health (NIH) establishes the Recombinant DNA Advisory Committee to provide federal oversight for all recombinant DNA research. The committee continues to function today as a forum for scientific, ethical, and legal issues raised by recombinant DNA technology and its research and clinical applications.

German-American biochemist Rudolf Jaenisch produces the first transgenic animal when he injects a cancer gene into mouse embryos. At maturity, the mice born from this experiment carry the cancer gene on their own DNA. The gene is transmitted, along with the rest of the mouse genome, to succeeding generations.

1975   The Asilomar Conference on Recombinant DNA is held at the Asilomar (California) Conference Center. About 140 scientists, physicians, and legal experts meet to draw up voluntary guidelines for the conduct of recombinant DNA research because of its potential for serious health and ethical issues.

1976   American venture capitalist Robert A. Swanson and Stanley N. Cohen cofound Genentech, Inc., a corporation designed to develop and market applications of rDNA technology originally discovered by Cohen and Boyer in 1973.

The NIH issues the first guidelines for rDNA research. The guidelines are based largely on recommendations from the Asilomar Conference (1975).

1977   Genentech announces the production of the first human protein manufactured by bacteria—somatostatin—a human growth hormone. The protein is the first commercial product of industrial rDNA research.

1980
The U.S. Supreme Court rules that a patent may be issued to the General Electric company for a genetically engineered bacterium invented by Indian-American chemist Ananda Chakrabarty. The court rules that the fact that the item in question is alive is not an impediment to its being patented. The decision sets a precedent for the granting of patents for all manner of engineered organisms and components of life (such as genes).

A group of Canadian researchers successfully introduces the human gene for the production of the protein interferon into a bacterium, the first time a human gene has been so transferred. The discovery makes possible the relatively simple and inexpensive production of the medically valuable protein.

American biochemist Kary Mullis invents the polymerase chain reaction (PCR), a method for making a very large number of copies of a given DNA specimen. PCR is now one of the most widely used procedures in DNA technology.

The U.S. Patent and Technology Office issues patents to Stanford University for the fundamental technology used in recombinant DNA technology. Over ensuing years, Stanford sells licenses to at least six dozen companies for the use of this technology in their own research and development programs.

1982
California researcher Stephen Lindow requests permission from the NIH to field test an engineered bacterium that promises to make strawberries, potatoes, and other plants less vulnerable to frost damage. A year later, the NIH granted permission for the tests.

1983
Researchers locate the genetic marker for Huntington's disease on chromosome number 4, the first

time a disease-causing gene has been specifically mapped in the human genome.

1984   British biochemist Sir Alec Jeffreys develops a process known as restriction fragment length polymorphism (RFLP), a technique by which DNA from different organisms can be distinguished from each other. It later becomes a powerful tool in forensic analysis.

Researchers at the University of Pennsylvania, the U.S. Department of Agriculture Agricultural Research Service, and the University of Washington create transgenic pigs, sheep, and rabbits by injecting foreign DNA into the organisms' eggs and allowing them to develop and come to term.

1985   The NIH issues the first guidelines for human gene therapy.

1986   RFLP analysis (see 1984) is used to convict Colin Pitchfork of the murder of two girls in Narborough, England. The case is the first incident in which DNA evidence is used to solve a crime.

The U.S. Department of Agriculture (USDA) authorizes the first release of a genetically engineered organism, a type of tobacco plant developed by the Agracetus company that is resistant to crown gall disease.

The U.S. Food and Drug Administration (FDA) grants approval for use of the first genetically engineered vaccine, a product made by the Chiron company for use against the hepatitis-B virus.

All agencies of the federal government responsible for the oversight of DNA technology in the United States join to issue a set of guidelines, *Coordinated Framework for Regulation of Biotechnology*, that clearly specifies the responsibilities assigned to the FDA, NIH, U.S. Environmental Protection

1986 (*cont.*)	Agency (EPA), USDA, and other agencies involved in regulation.

For the first time, researchers clone a disease-producing gene, the gene that causes chronic granulomatous disease, using a method known as positional cloning. Positional cloning allows one to find a gene in the human genome without knowing anything about the protein it produces.

1987	A committee especially chartered by the U.S. Congress, the Health and Environmental Research Advisory Committee (HERAC), recommends the creation of a 15-year program to sequence the human genome (see 1989).

1988	The USDA authorizes field tests of the Calgene corporation's Flavr Savr™ genetically engineered tomato. The tomato carries a gene that suppresses the gene that causes a plant to continue ripening. The product is later approved for production and sale, but economic issues cause the company to discontinue its production in 1997.

Harvard molecular geneticists Philip Leder and Timothy Stewart are awarded the first patent for a genetically altered animal, a mouse that is highly susceptible to breast cancer, the so-called Harvard Oncomouse. The European Union later approves a patent for the same discovery in 2001, although the Canadian Supreme Court denies a similar application in 2002.

The U.S. Federal Bureau of Investigation (FBI) establishes a laboratory for the testing of DNA samples in criminal investigations.

Tommy Lee Andrews, a resident of Orlando, Florida, is convicted of rape at least partly as the result of DNA typing evidence submitted by the prosecution. The case is the first instance in the United State in which DNA evidence is accepted in a criminal case.

On the basis of DNA evidence obtained from the scene of the crime, Gary Dotson is exonerated of rape and aggravated kidnaping charges for which he was convicted in 1979.

1989   The National Center for Human Genome Research is established for the purpose of conducting and sponsoring research on the human genome and related issues. The first director of the center is James Watson.

The admissibility of DNA evidence is challenged for the first time in the United States in the case of *People v. Castro*. The court rules that DNA evidence can be used to exclude suspects, but not as evidence of guilt.

The Virginia Supreme Court confirms the death penalty for Timothy Wilson based on DNA evidence presented in his earlier trials. The case is the first instance in the United States in which DNA evidence leads to the death penalty.

1990   Researchers at the National Heart, Lung, and Blood Institute and the National Cancer Institute perform the world's first human gene therapy treatment on two children with adenosine deaminase deficiency, a severe hereditary disease. Later studies appear to show that the treatment is at least partially successful, and a similar procedure has since been used with more than two dozen patients worldwide.

The GenPharm corporation uses the world's first transgenic dairy cows to produce a component of human milk that can be incorporated into infant formula to make it more nutritious and more like human milk.

The biotechnology company Calgene develops an engineered form of cotton resistant to the herbicide bromoxynil (Buctril®), used to treat broadleaf weeds in cotton (and other crop) fields.

The FBI initiates a pilot DNA database program called the Combined DNA Index System (CODIS), designed to provide an efficient system by which federal, regional, state, and local law enforcement agencies can exchange DNA typing evidence from criminal scenes.

1992   The USDA issues a policy statement indicating that food from genetically engineered plants will not be regulated any differently from conventional foods.

A team of British and American doctors develops a procedure for testing embryos for the presence of certain genetic disorders, such as hemophilia and cystic fibrosis. They remove individual cells from four- or eight-cell embryos and look for defective genes that may cause such disorders.

1993   The Biotechnology Industry Organization is formed by the merger of the Association of Biotechnology Companies and the Industrial Biotechnology Association.

The FDA approves the use of genetically engineered bovine somatotropin hormone (bST) for use in dairy cows to increase milk output.

Kirk Bloodsworth is exonerated by DNA typing of the 1984 murder and rape of a nine-year-old girl in Rosedale, Maryland. He is the first person on death row to be exonerated as a result of the technology.

1994   The FDA approves the marketing of the Flavr Savr™ tomato.

1995 The EPA approves the first plant genetically engineered to resist attack by pests, the New Leaf potato, developed by Monsanto.

    The Ethical, Legal, and Social Issues (ELSI) division of the Human Genome Project (HGP) publishes a model Genetic Privacy Act recommended for adoption by individual states until the U.S. Congress enacts a federal law dealing with privacy issues occasioned by the use of genetic testing.

    A team of researchers from Johns Hopkins University, the State University of New York in Buffalo, the National Institute of Standards and Technology, and the Institute for Genomic Research sequence the first complete genome of a free-living organism, the bacterium *Hemophilus influenzae.* The genome consists of 1,749 genes.

1996 Initial research in the sequencing of the human genome under the Human Genome Project begins at six universities in programs sponsored by the federal government.

    A joint NIH/Department of Energy Committee appointed to evaluate the ELSI program of the Human Genome Project issues a report containing recommendations about future directions of the ELSI component of the HGP.

    The Office of Justice Programs at the U.S. Department of Justice publishes a report, *Convicted by Juries, Exonerated by Science: Case Studies. in the Use of DNA Evidence to Establish Innocence After Trial,* that summarizes the use of DNA typing in proving the innocence of men and women previously convicted of crimes.

1997 The National Center for Human Genome Research is promoted to full institute status within the NIH, and its name is correspondingly changed to the National Human Genome Research Institute.

Researchers at the Human Genome Project begin to identify specific genes responsible in full or part for a variety of human diseases, including breast cancer, Parkinson's disease, and Pendred Syndrome.

HGP researchers completely sequence the first human chromosome, chromosome number 7.

The European Parliament adopts a set of rules and regulations concerning the production and sale of genetically modified foods within the European Union.

The NIH-Department of Energy Working Group on Ethical, Legal, and Social Implications of Human Genome Research issues a report on genetic testing, *Promoting Safe and Effective Genetic Testing in the United States: Final Report of the Task Force on Genetic Testing*, that makes no recommendations about specific tests, but suggests some general guidelines to ensure that tests are conducted and interpreted properly.

Researchers at the Roslyn Institute in Edinburgh, Scotland, clone the first mammal, a sheep given the name of Dolly, from an adult somatic cell by the process of nuclear transfer.

The Monsanto corporation introduces the first herbicide-resistant food crop, Roundup Ready® soybeans, and the first pesticide-resistant crop, Bollgard® cotton.

1999   Eighteen-year-old Jesse Gelsinger is the first person reported to have died following a gene therapy procedure. Gelsinger was being treated for ornithine transcarbamylase deficiency, a condition that is normally fatal at birth, but which he had survived because of its occurrence as the result of a mutation rather than a genetic defect.

2000   President Bill Clinton issues Executive Order 13145, which prohibits discrimination in federal employment based on genetic information.

Ingo Potrykus, emeritus professor of plant sciences at the Swiss Federal Institute of Technology, and his colleagues announce the availability of golden rice, rice that has been genetically engineered to produce beta-carotene, used in the production of vitamin A. The new form of rice provides a simple and inexpensive way to prevent blindness in millions of children around the world.

2001   The FDA issues a set of guidelines for the voluntary labeling of foods produced in the United States that have been produced by some type of genetic engineering. Lacking federal legislation on this issue, these guidelines are the closest thing currently available to legal regulations on the labeling of genetically modified foods.

Two separate research groups announce the nearly complete sequencing of the human genome. One group consists of researchers in the Human Genome Project, and the second group is a privately funded group of researchers at Celera Genomics. The "nearly complete" genome consists of about 83 percent of all known human genes.

The FDA approves the use of the drug imatinib (Gleevec®) for the treatment of certain types of cancer. Imatinib is the first of a group of so-called "targeted drugs"—drugs that act on one specific gene to prevent the production of a harmful protein/enzyme responsible for a disease. The ability to make targeted drugs is a result of scientists' new understanding of the human genome.

Researchers at Texas A&M University clone the first pet, a kitten named C.C. (for carbon copy).

2001 (*cont.*)	At a cost of about $50,000, the pet owner received an exact copy of a favorite cat who had died. The U.S. Equal Employment Opportunity Commission (EEOC) files suit against the Burlington Northern Santa Fe Railroad to end genetic testing of some of its employees without their knowledge. The suit is settled out of court a year later. The case was the first instance in which the EEOC filed suit against genetic testing based on the American with Disabilities Act of 1990.
2002	The USDA creates a new agency within the Animal and Plant Health Inspection Service, the Biotechnology Regulatory Services. The mission of the service is to focus on the department's responsibilities for regulating and facilitating biotechnology. The Human Genome Project announces the creation of the International HapMap Project, a program that allows access to information so that researchers can pursue studies on the relationship of specific genes in the human genome responsible for a number of common diseases, including asthma, cancer, diabetes, and heart disease. Merck & Company develops a genetically engineered vaccine for use against four types of human papillomavirus (HPV), implicated in the development of cervical cancer. The vaccine is developed by recombinant DNA technology. Researchers at the State University of New York at Stony Brook synthesize the first artificial virus, an identical copy of the virus that causes polio. They use easily obtained information and materials and call the process "very easy to do," although their work raises serious ethical concerns about the use of the technique by terrorists and the possibility of using it to create higher forms of life.

A team of British scientists discovers a gene that increases one's susceptibility to depression, raising the possibility of treating mental disorders with genetic therapy.

China's State Food and Drug Administration grants the world's first regulatory approval of a drug called Gendicine for the treatment of squamous cell head and neck cancer. The drug contains an engineered copy of the p53 gene, known for its ability to destroy cancer cells.

Researchers at the University of Georgia clone a cow from cells taken from a dead carcass. The method promises to improve the quality of meats since scientists can take samples from steaks and other meats of special quality and use them to clone animals with similar meat properties.

Researchers at the University of North Carolina develop a method for treating the genetic disorder thalassemia by repairing erroneous copies of messenger RNA rather than attempting to repair the incorrect gene responsible for the production of the mRNA.

Researchers at Emory University in Atlanta cure mice of sickle-cell anemia by transplanting corrected copies of defective genes responsible for the disease into the animals.

The National Academy of Sciences publishes a report on the environmental effects of transgenic plants, which includes a number of recommendations on the topic. The investigating committee points out that the environmental impacts of conventional farming are poorly known and should be studied, along with similar research on transgenic plants—such plants should be studied and evaluated on a case-by-case basis.

2003   Attorney General John Ashcroft announces creation of the President's DNA Initiative, a program designed to ensure that DNA typing is used to its greatest potential in solving crimes, protecting the innocent, and identifying missing persons. The initiative includes funding, training, and assistance for federal, state, and local forensic laboratories and law enforcement personnel.

In February, the FDA places a temporary hold on all human gene therapy trials that use retroviral vectors to insert genes. The ruling comes when the second of two French children treated (successfully) for X-linked severe combined immunodeficiency disease (X-SCID) develops leukemia. The FDA lifts the ban in April 2003 when it becomes clear that the technology used in the French experiments is not fundamentally flawed.

2004   A team of Japanese and American researchers develops methods for producing a transgenic animal (zebrafish), resulting from the fertilizing of conventional fish eggs by sperm that have been genetically modified in a laboratory dish. The technology holds significant progress for the production of other transgenic animals by methods that are qualitatively different from those used in the past.

The United Nations Food and Agriculture Organization issues a report, *The State of World Food and Agriculture 2004*, that says that genetically engineered crops may provide a huge benefit to the millions of people around the world who do not get enough to eat every day, but that questions of the risks to human health and the environment from such crops remain to be answered.

The U.S. Institute of Medicine issues a report that says that genetically modified foods pose no more health risks than foods produced by conventional methods. The report suggests that all foods should be evaluated on the basis of their properties and characteristics, and not on the basis of the method used to produce them.

The FDA approves an application by the company Genomic Health for one of its products, Oncotype DX™, for the determination of the likelihood of reoccurrence of breast cancer. The test measures the activity of certain genes known to be implicated in the development of breast cancer.

Voters in California approve Proposition 69, which requires that all convicted felons and anyone arrested for or charged with felonies and some misdemeanors be required to submit DNA samples for placement in a state DNA database. The proposition applies to both adults and juveniles. The proposition passes 62 percent to 38 percent.

2005   The United Nations General Assembly adopts a somewhat ambiguous resolution opposing cloning for the purposes of producing human life but, presumably, supporting cloning for therapeutic reasons.

The World Health Organization (WHO) issues a report, *Modern Food Biotechnology, Human Health and Development*, which concludes that engineered crops can increase yield, improve food quality, and expand the diversity of foods available to consumers, leading to improved nutrition and better overall health for millions of people around the world.

2006   The NIH announces a new project called the Genes and Environment Initiative with two objectives: (1) discovering genetic differences among individuals with the same disease; and (2) finding ways to identify the interaction between heredity and environment in the development of specific diseases.

2006
(*cont.*)

A public-private partnership called the Genetic Association Information Network is created to study the genetic components of six major diseases: attention deficit hyperactivity disorder, bipolar disorder, diabetic nephropathy, major depressive disorder, psoriasis, and schizophrenia.

The USDA's Center for Veterinary Biologics gives approval to Dow AgroSciences for the production of the world's first plant-made vaccine. The vaccine is used to protect chickens from New castle disease, a highly contagious and fatal viral infection. The production process for the vaccine makes use of plant parts only, and not complete plants, and is conducted in an enclosed facility to prevent escape of genetic components into the environment.

The USDA gives its approval to the Renessen Corporation to begin manufacture and sale of its genetically engineered corn Mavera™, which contains an added gene for the production of lysine. Mavera™ corn is more nutritious for swine and poultry than is conventional corn, improving the quality of meat and reducing the cost of production for dairy operators.

The World Trade Organization rules that the European Union's ban on genetically modified foods is illegal in light of overwhelming evidence that engineered foods pose no more of a risk to human health than do conventional foods. The ruling comes in response to a complaint filed by Argentina, Australia, Brazil, Canada, India, Mexico, New Zealand, and the United States.

Researchers at the National Cancer Institute use genetically engineered immune cells to successfully treat two patients with metastatic melanoma, an aggressive form of cancer. Fifteen other patients in the study receive no benefit from the treatment, but the study confirms the possibility of using gene therapy for the treatment of at least some forms of cancer.

2007 British scientists use gene therapy to successfully treat patients with Leber's congenital amaurosis, a genetic disorder that causes severe loss of sight and/or blindness.

2008 The U.S. Congress passes and President George W. Bush signs the Genetic Nondiscrimination Act, which prohibits discrimination in hiring, employment, and related situations on the basis of information obtained from genetic testing.

The United Nations adopts an additional protocol to its Convention on Human Rights and Biomedicine opposing discrimination based on information obtained as the result of genetic testing.

The FDA approves the use of Recothrom®, a genetically engineered product that induces blood clotting, with potentially extensive use in surgery when typical blood-clotting cannot be used.

2009 The FDA announces its final ruling for the regulation of genetically engineered animals. Such animals are to be approved on a case-by-case basis by the FDA, but food products produced from them do not have to carry special labels.

Researchers at an Israeli forensic science laboratory announce that they have found that it is relatively easy to falsify DNA evidence collected at a crime scene using simple equipment and understanding of DNA typing procedures.

5

Biographical Sketches

This chapter contains brief biographical sketches of some individuals who have played an important role in the development of DNA technology. The names included here represent only a small fraction of the men and women who made basic discoveries about DNA and its chemical, physical, and biological properties and about its many applications in forensic science, human gene therapy, genetically modified organisms, genetic testing, and other fields.

W. French Anderson (b. 1936)

Anderson is sometimes called the father of human gene therapy because of his role in some of the earliest efforts to cure disease by gene transfer methods. In 1990, he treated a four-year-old girl with severe combined immunodeficiency disease (SCID), which occurs when a genetic error prevents a child's immune system from developing normally. As a result, the child is highly susceptible to all forms of infection and, often, dies at an early age. Anderson and his colleagues proposed to cure the disease by introducing functioning copies of the defective genes into the child's genome, restoring its ability to produce healthy components of the immune system.

William French Anderson was born in Tulsa, Oklahoma, on December 31, 1936. He developed an avid interest in science and mathematics at an early age, but was also interested in a variety of extracurricular activities, including theater, track, and debate. He is said to have become interested in the last of these subjects because of his tendency to stutter, a problem he thought he could resolve with demanding attention to speaking. He also later expressed an interest in studying the genetic basis for stuttering.

Anderson entered Harvard College in 1954, where he was exposed to the most recent advances in genetics, especially in a course taught by James Watson. The course was instrumental in convincing him to change his college major from mathematics to genetics. After receiving his B.S. from Harvard in 1958, he went to England to continue his studies under Francis Crick at Cambridge University. He earned his M.A. from Cambridge in 1960, and then returned to Harvard to study medicine, earning his M.D. in 1963.

In 1965, Anderson took a research position with the National Institutes of Health (NIH), where he spent the next 30 years working on methods for treating genetic disorders with engineered cells. He chose to focus on a disease known as adenosine deaminase deficiency disease (ADA) because it develops as the result of a single defective gene, a much simpler condition than many other types of genetic disease. ADA is one of the conditions commonly associated with SCID. Anderson extracted T cells from the four-year-old patient, inserted correct copies of the ADA gene into those cells, and then returned the engineered cells to her body. He was able to show that the cells survived, reproduced, and began to synthesize the adenosine deaminase previously absent from her immune system.

In 1992, Anderson left the NIH to accept an appointment as professor of biochemistry and pediatrics and director of the Gene Therapy Laboratories at the Keck School of Medicine at the University of Southern California. In 2004, he was charged with sexual abuse of a minor, convicted of the offense, and sentenced to 14 years in prison.

Werner Arber (b. 1929)

Arber discovered an important family of chemical compounds known as *restriction enzymes* that cut DNA molecules at specific points in their polynucleotide chains. Prior to Arber's work, scientists had found that bacteria have evolved mechanisms for protecting themselves against attacks by viruses (called *bacterio-phages*, or just *phages*), just as higher organisms have evolved immune systems to protect themselves against attack by viruses, bacteria, fungi, and other foreign organisms. Arber was able to show that the "immune system" used by bacteria consists of enzymes that cut the DNA of invading viruses, disabling them and preventing their continued infection of the bacteria. Arber's discovery was an important milestone in the development of DNA technologies, because it provided a method by which DNA molecules can be taken apart for the insertion of gene fragments from other sources, making possible the formation of recombinant DNA molecules.

Werner Arber was born in Gränichen, Switzerland, on June 3, 1929. After completing his public school education in 1945, he enrolled at the Kantonsschule Aarau gymnasium, from which he received a second-level "maturity" certificate in 1949. He continued his studies in chemistry and physics at the Swiss Polytechnical School in Zürich, where he earned his degree in 1953. Arber then chose to accept an assistantship in electron microscopy at the University of Geneva, a position in which he became more interested in and better informed about problems of bacteriophage physiology and genetics. He eventually decided to make a study of phages the focus of his career.

Arber was awarded his doctorate at Geneva in 1958 and then spent two years in a postdoctoral program at the University of Southern California in phage genetics. Towards the end of that program, he had an opportunity to visit the laboratories of a number of pioneers in that field, including the laboratories of Gunther Stent, Joshua Lederberg, and Salvador Luria, where he learned about the most recent advances in phage genetics and molecular biology. When Arber returned to the University of Geneva in 1960, he began the research that would lead to his discovery of restriction enzymes.

Arber was promoted to associate professor in molecular biology at Geneva in 1965 and, in 1971, he moved to the University of Basel as full professor of molecular microbiology, a post he held until his retirement in 1996. From 1986 to 1988, Arber also served as rector of the University of Basel. He was awarded a share of the 1978 Nobel Prize for Physiology or Medicine. Arber continues his affiliation with the University of Basel, where he is emeritus professor of molecular microbiology.

Oswald Avery (1877–1955)

In 1944, Avery reported on a classic series of experiments conducted with postdoctoral colleagues Colin MacLeod and Macyln McCarty over a period of years. Those experiments were designed to identify the unit within living organisms responsible for the transmission of genetic traits. At the time, most scientists thought that some type of protein played this role. Heredity is a complex phenomenon that requires a complex molecule for transmission of traits to occur. Proteins certainly fit that role as they are large, complex molecules. In the Avery-MacLeod-McCarty experiments, laboratory mice were injected with two forms of pneumonococci (the bacteria that cause pneumonia), one that was virulent (disease-causing and fatal) and one that was not. By treating the mice with various combinations of bacteria, the researchers found that the compound responsible for transmitting resistance to pneumonococci in mice was DNA, not a protein. This research is of such profound significance in the history of DNA that Avery is sometimes described as the most deserving researcher never to have received a Nobel Prize.

Oswald Theodore Avery was born on October 21, 1877, in Halifax, Nova Scotia. He moved with his family to New York City in 1887, where he attended the New York Male Grammar School. He continued his studies at the Colgate Academy, preparatory school of Colgate University, and then at the university itself. He earned his B.A. in humanities in 1900, but then chose to study medicine at the Columbia University College of Physicians and Surgeons in New York, where he received his M.D. in 1904.

Avery remained in practice for only a relatively short time, primarily because he was frustrated at his inability to help some of his patients. He decided instead to turn to a career in research and, in 1907, accepted a position as associate director of the division of bacteriology at the Hoagland Laboratory in Brooklyn, New York. Of the many research papers he wrote at the Hoagland, one caught the eye of Rufus Cole, then director of the Hospital of the Rockefeller Institute for Medical Research. Impressed by Avery's work, Cole offered him a position at Rockefeller in 1913. He remained at Rockefeller until he retired from research in 1948. Although he was given emeritus status by the university in 1943, Avery continued his research there for another five years. It was near the end of his active involvement in scientific research, then, that he, along with MacLeod and McCarty, made his revolutionary discovery about DNA. In 1948, Avery left Rockefeller and moved to Nashville, Tennessee, to be near his brother. He died in Nashville on February 2, 1955, of pancreatic cancer.

Paul Berg (b. 1926)

In the late 1960s, Berg made the first recombinant DNA (rDNA) molecules by splicing together foreign genes from a variety of sources into the DNA molecules of host organisms, such as bacteria and mammalian cells. This achievement not only brought together a number of techniques previously developed to produce recombinant organisms, but also provided a method for studying the characteristics and outputs of specific segments of DNA, those inserted into host organisms.

Paul Berg was born in New York City on June 30, 1926. He attended Pennsylvania State University, from which he received his B.S. in biology in 1948, and Western Reserve University (now Case Western Reserve University), where he was awarded his Ph.D. in biochemistry in 1952. After four years of postdoctoral research at the NIH, the Institute of Cytophysiology in Copenhagen, and the Washington University School of Medicine in St. Louis, Berg accepted an appointment as assistant professor of microbiology at the Washington University School of Medicine. In 1959, Berg moved to Stanford University as associate professor in the School of Medicine. He spent the rest of his academic career at Stanford in a number of positions, including Sam, Lulu, and Jack Willson Professor of Biochemistry; Vivian K. and Robert W. Cahill Professor in Biochemistry and Cancer Research; Director, Beckman Center for Molecular and Genetic Medicine; and (currently) Cahill Professor in Biochemistry, Emeritus, and Director of the Beckman Center for Molecular and Genetic Medicine, Emeritus. He was elected to the National Academy of Sciences in 1966 and was awarded the Nobel Prize in Chemistry in 1980 for his research on recombinant DNA molecules.

After announcing the results of his research on recombinant DNA, Berg became concerned about possible safety, social, and ethical implications of such research. He wrote to the NIH, outlining these concerns, and encouraged his fellow scientists to meet and consider the possible risks involved in rDNA research. Largely because of Berg's efforts, a group of more than 100 researchers, legal experts, and others concerned about rDNA research met at the Asilomar Conference Center in California in 1975 to work out guidelines for the monitoring and control of such research. Those guidelines were the first regulations ever developed for research on DNA technology.

Herbert W. Boyer (b. 1936)

In the history of DNA technology, the names of Herbert Boyer and Stanley N. Cohen will forever be paired with each other. In 1972, the two researchers were both attending a conference in Hawaii on plasmids, circular loops of DNA found in bacteria and protozoa. Over lunch, the two men found that they were engaged in very similar research projects. Cohen was studying the antibiotic properties of certain bacterial plasmids, while Boyer was studying methods for introducing DNA into precisely defined segments with ends that could be attached to other pieces of DNA. The two decided to collaborate, and within four months, had carried out one of the classic studies in the history of molecular biology. In this study, they introduced specified pieces of DNA into a bacterial plasmid (using methods developed by Boyer), and then inserted the plasmid into bacteria (using methods developed by Cohen). The results of these experiments were bacteria whose DNA contained clearly defined segments of foreign DNA (genes) capable of synthesizing specific proteins. When those bacteria reproduced, they then became tiny "factories" for the production of those proteins, the earliest forerunners of contemporary industrial rDNA technologies.

Herbert Wayne Boyer was born in Pittsburgh, Pennsylvania, on July 10, 1936. He received his A.B. from St. Francis College in 1968, his M.S. and Ph.D. (bacteriology) from the University of Pittsburgh in 1960 and 1963, respectively. He did his postdoctoral work at Yale University as a U.S. Public Health Service research fellow. In 1966, Boyer left Yale to accept a position at the University of California at San Francisco (UCSF) as assistant professor of microbiology. He has remained with UCSF ever since and currently holds the title of emeritus professor of biochemistry and biophysics in the UCSF School of Medicine.

In 1976, Boyer and venture capitalist Robert A. Swanson founded the world's first corporation for the development of commercial products made with rDNA research, Genentech, Inc. Within its first year of operation, Genentech had produced the first commercial rDNA product, the hormone somatostatin. Among the honors given to Boyer include the National Medal of Technology, National Medal of Science, Biotechnology Heritage Award, Shaw Prize in Life Science and Medicine, Lemelson-MIT Prize, and the Albany Medical Prize. The graduate faculty at UCSF has created the Herbert W. Boyer Program in Biological Sciences in recognition of Boyer's contribution to the university.

Erwin Chargaff (1905–2002)

Chargaff is best known for his discovery of the relationship among nitrogen bases in a DNA molecule, a relationship now known as Chargaff's Rules. According to those rules, the number of adenine units in a DNA molecule is equal to the number of thymine units, and the number of cytosine units is equal to the number of guanine units. A second, and somewhat less well-known rule, is that the composition of DNA (that is, the relative amounts of adenine, cytosine, guanine, and thymine) varies from organism to organism. Although Chargaff himself never realized the implications of his rules for the composition of the DNA molecule, those rules proved to be an essential clue for the elucidation of the structure of the DNA molecule by James Watson and Francis Crick in the early 1950s.

Erwin Chargaff was born in Czernowitz, Austria (now Chernovtsy, Ukraine), on August 11, 1905. He showed an early precocity in the study of languages (he eventually was able to speak and understand 15 languages) and considered a career in philology. He eventually settled on science, however, and attended the University of Vienna to pursue a degree in chemistry. He received his Ph.D. from Vienna in 1928 and then accepted a position at Yale University as Milton Campbell Research Fellow in Organic Chemistry. Two years later he returned to Europe where he worked as assistant in charge of chemistry for the Department of Bacteriology and Public Health at the University of Berlin and as research associate at the Institut Pasteur in Paris. In 1935 he returned again to the United States, where he became research associate in the Department of Biochemistry at Columbia University, where he remained for the rest of his academic career. He retired from Columbia in 1974 as professor emeritus. His retirement was accompanied by some ill feelings between Chargaff and the university, and he transferred his research activities to Columbia's affiliate, Roosevelt Hospital, where he continued to work until 1992. He died in New York City on June 20, 2002 at the age of 96.

By the 1950s, Chargaff had become somewhat disenchanted with the direction being taken by molecular biologists. He said that researchers were "running riot and doing things that can never be justified." He believed that nature was far too complex for humans to think of it and treat it as some sort of machine that can be manipulated at their will. He even warned that the manipulation of genetic materials was an even greater threat to the human race than was nuclear power.

Chargaff received a number of honors and awards during his lifetime, including the Pasteur Medal, the Carl Neuberg Medal, the Société de Chimie Biologique Medal, the Charles Leopold Mayer Prize, the H.P. Heineken Prize, the Bertner Foundation Award, the Gregor Mendel Medal, the National Medal of Science, and the New York Academy of Medicine Medal.

Mary-Dell Chilton (b. 1939)

Chilton is best known for her research on the process by which *Agrobacterium* bacteria infect tobacco plants. Her team showed that a bacterium is able to transmit its DNA into the host plant genome, and that, furthermore, removal of disease-causing genes from the bacterial DNA does not affect this process. The information gained from these studies demonstrated the feasibility of producing transgenic plants by inserting foreign genes into the genome of a host plant, giving it the ability to produce proteins not typically associated with the native plant. The experiments conducted by Chilton's team have now become classics in the field of genetic engineering of plants.

Mary-Dell Chilton was born in Indianapolis, Indiana, on February 2, 1939. She was originally interested in astronomy and was a finalist in the 1956 Westinghouse Science Talent Search for building "a long telescope in a short tube." She planned to major in astronomy when she entered the University of Illinois at Champaign Urbana in 1956, but soon found that her instructors did not take her seriously because she was a woman. She briefly transferred to physics, but found that subject boring, so eventually decided to major in chemistry, earning her B.S. in that subject in 1960 and her Ph.D., also in chemistry, in 1967. Chilton then accepted an appointment at the University of Washington, in Seattle, where she remained until 1979. It was at the University of Washington that she conducted much of her initial work on the production of transgenic plants.

In 1979, Chilton moved to Washington University in St. Louis, where she continued her research on the genetic modification of tobacco plants. After four years, she resigned her academic position in order to take a job with the CIBA-Geigy Corporation (now Syngenta Biotechnology, Inc.) in Research Triangle Park, North Carolina. At Syngenta, she has been involved in both research and administrative activities, serving as principal scientist, distinguished science fellow, and vice president of agricultural biotechnology. In 2002, she received one of the highest awards in her field, the Benjamin Franklin Medal in Life Sciences of the Franklin Institute.

Stanley N. Cohen (b. 1935)

Cohen and Herbert Boyer performed one of the classic studies in the early history of DNA technology in the early 1970s when they found a way to insert a foreign gene into a plasmid, a double-stranded circular piece of DNA found in bacteria and protozoa. This technology allowed Cohen and Boyer to make duplicate copies (clones) of precise segments of DNA from any given source.

Stanley Norman Cohen was born in Perth Amboy, New Jersey, on February 17, 1935. He attended Rutgers University, from which he received his B.A. in biological sciences in 1956, and the University of Pennsylvania School of Medicine, from which he received his M.D. in 1960. He then did his residency, internship, and medical research at a number of institutions, including University Hospital at the University of Michigan, Mt. Sinai Hospital in New York City, Duke University Hospital, and the National Institute of Arthritis and Metabolic Diseases. In 1968, Cohen was appointed assistant professor of medicine at Stanford University, where he spent the remainder of his academic career.

He is currently professor of genetics and professor of medicine at the Stanford School of Medicine.

Cohen and Boyer met at a conference in Hawaii on bacterial plasma and over lunch discovered that their research interests melded with each other beautifully. At the time, Cohen was studying methods for inserting plasmids into bacteria in order to study their ability to develop resistance to certain antibiotics. Boyer was working on the development of certain kinds of enzymes that cut DNA into precisely defined segments with "sticky" ends. The two combined their skills to develop a method for inserting precise DNA segments into plasmids and then inserting those plasmids into bacteria. The modified bacteria could, by this method, be "engineered" to produce any desired protein product specified by the inserted DNA and, in this regard, marked the beginning of industrial biotechnology based on rDNA molecules.

In his long and illustrious career, Cohen has received a number of awards and honors, including the National Medal of Science, the Lemelson-MIT Prize, the National Biotechnology Award, and the Albert Lasker Basic Medical Research Award.

Sir Francis Crick (1916–2004)

Crick received a share of the 1962 Nobel Prize for Physiology or Medicine for his discovery, with James Watson, of the molecular structure of the DNA molecule. This research marked a turning point in the history of genetics, since it clarified for the first time the precise nature of a gene. Having long been thought of as some ambiguous "black box" by which hereditary characteristics are passed from one generation to the next, the gene was now describable as a discrete chemical molecule whose structure and behavior could be analyzed, understood, and, at least in theory, manipulated. Almost all forms of DNA technology would be impossible without this information.

Francis Harry Compton Crick was born in Weston Favell, England, on June 8, 1916. He received his B.S. in physics at University College, London, in 1937 and then began a Ph.D. program in physics under the eminent E. N. da C. Andrade. That work was interrupted by World War II, during which time Crick served as a researcher for the British Admiralty on magnetic and acoustic mines. In 1947, he left the Admiralty and returned to his studies, but this time in the field of biology. Like many physical scientists of the time, Crick had become fascinated by the puzzles posed by biological phenomena and the increasing power of chemical and physical technologies for solving these problems. He eventually was assigned to the Cavendish Laboratory at the University of Cambridge, where he came under the influence of Sir Lawrence Bragg, one of the founders of the science of X-ray crystallography. X-ray crystallography is a technique for using X rays to elucidate the structure of complex molecules, like proteins and nucleic acids. In 1954 he received his Ph.D. for the analysis of polypeptides and proteins with X rays.

In 1951, Crick met James Watson, a postdoctoral student from the United States, at the Cavendish. The two men found that they had many common interests and decided to collaborate on efforts to determine the molecular structure of DNA. They succeeded in that effort in late 1952 and published the results of their work in the following year.

In the years following the discovery of DNA structure, Crick turned his attention to a number of other topics, including the process by which proteins are synthesized using the code stored in DNA molecules, the nature of the DNA code itself, the origins of life, and the chemical structure and properties of the nervous system and the brain. In 1976, Crick left Cambridge to take a position at the Salk Institute for Biological Studies in La Jolla, California. Later, he was also appointed to the faculty at the University of California at San Diego. He received many awards and prizes in addition to the Nobel Prize and was knighted by Queen Elizabeth II in 2002. Crick died in San Diego of colon cancer on July 28, 2004.

Ronald G. Crystal (b. 1941)

In 1993, Crystal led a team of researchers who performed the first human gene therapy on a patient with cystic fibrosis, using a genetically engineered live adenovirus. The virus had also been modified to eliminate its ability to self-replicate, thereby eliminating the possibility that it would reproduce within the patient's body or the surrounding environment. The experiment was only a partial success as the patient's immune system soon responded to the foreign virus and destroyed it, making it impossible for the corrective genes it carried to begin functioning. The experiment was a success, however, in that it validated the method by which genes can be safely introduced into a patient's body in an effort to correct a genetic disorder.

Ronald George Crystal was born in Newark, New Jersey, on April 23, 1941. He received his B.A. in Physics from Tufts University, in Medford, Massachusetts, in 1963, his M.S. in physics and his M.D. at the University of Pennsylvania in 1963 and 1968, respectively. He then completed his residency and internship at Massachusetts General Hospital, in Boston, from 1968 through 1970. In 1970, he joined the National Heart and Lung Institute (now the National Heart, Lung, and Blood Institute; NHLBI) of the National Institutes of Health as a research associate in the section on molecular hematology. In the same year, he was commissioned as a surgeon in the Commissioned Corps of the U.S. Public Health Service (USPHS). He has continued his affiliation with the Commissioned Corps ever since, now holding the title of medical director (retired) of the corps. He also retained his association with NHLBI, where he was chief of the pulmonary branch from 1975 to 1993.

In 1993, Crystal joined the Weill Medical College of Cornell University, where he held concurrent posts as Bruce Webster Professor of Internal Medicine, chief of the Division of Pulmonary and Critical Care Medicine at New York Presbyterian Hospital, and professor of physiology and biophysics in the Weill Medical College Graduate School. He has since also been appointed as director of the Belfer Gene Therapy Core Facility, professor of genetic medicine, and chairman of the department of genetic medicine at Weill.

Rosalind Franklin (1920–1958)

Franklin was a biophysicist and X-ray crystallographer whose images of the DNA molecule provided invaluable information used by James Watson and Francis Crick in their elucidation of the structure of the molecule.

Rosalind Elsie Franklin was born at Notting Hill, London, on July 25, 1920 into a prominent and politically active Jewish family. Her father was Ellis Arthur Franklin, a successful banker who taught physics and history at the Working Men's College during the evening. After graduating from St. Paul's Girls School in London in 1938, Franklin enrolled at Newnham College at the

University of Cambridge. Three years later, she completed her course of study, but was granted only a titular degree since women were not granted the same B.A. as were men. She continued her studies at Cambridge and in 1941 was granted her Ph.D. in chemistry. During World War II, Franklin worked at the British Coal Utilisation Research Association, studying the properties of coal, a topic on which she eventually became an internationally-recognized authority.

When the war ended, Franklin accepted an appointment at the Laboratoire central des services chimiques de l'État in Paris, where she was first introduced to the use of X-ray crystallography for the analysis of complex structures. She remained in France until 1950, when she was offered a job as a research associate at King's College, London, under the famous British biophysicist John Randall. Her assignment there was to study the crystalline structure of DNA fibers.

Franklin's tenure at King's produced the most remarkable images of DNA yet produced, but her work there was suffused with controversy and confusion. By late 1952 she had, in fact, obtained images that made very clear the true structure of the DNA molecule, a fact that she was not quite willing to accept, but that James Watson and Francis Crick quickly understood. The shabby way in which Franklin was treated by her male colleagues at King's was never fully understood until the publication of Watson's story of the account of the DNA discovery in his *The Double Helix*, published in 1968. By that time, Franklin had been dead for a decade, having died on April 16, 1958, in Chelsea, London, of ovarian cancer. Today she is memorialized by the Rosalind Franklin University of Medicine and Science, formerly the Chicago Hospital College of Medicine.

Timothy C. Hall (b. 1937)

A research team headed by Hall and U.S. Department of Agriculture (USDA) biochemist John D. Kemp developed a method for transferring a gene from one plant, a bean, into a second plant, a sunflower, after which the bean gene began to express itself in the sunflower plant. The experiment provided one of the earliest pieces of evidence that the genome of plants can be altered by human interventions, the basic principle underlying the modern science of pharming.

Timothy Couzens Hall was born in Darlington, England, on August 29, 1937. He attended the University of Nottingham from 1959 to 1965, earning his Ph.D. in plant physiology there in 1965. After a year of postdoctoral research at the University of Minnesota, Hall took a position as assistant professor of horticulture at the University of Wisconsin, where he eventually became full professor in 1975. From 1980 to 1984, Hall served concurrently as a director of the Agrigenetics Research Corporation in Boulder, Colorado.

In 1984, Hall accepted an appointment as distinguished professor and head of biology at Texas A&M University, in College Station. He was later named director of the Institute of Developmental and Molecular Biology at Texas A&M, a post he continues to hold today. Hall holds five patents for methods of transporting and obtaining expression of foreign genes into host plants.

Sir Alec Jeffreys (b. 1950)

Jeffreys is discoverer of the restriction fragment length polymorphisms (RFLP) technique for the amplification of small segments of DNA, a key technology in a number of fields of DNA technology, including forensic science, the determination of paternity, and wildlife biology. RFLP has been replaced in many instances by the polymerase chain reaction (PCR) procedure developed by Kary Mullis, but it was the first such technology to be used and is still employed in some circumstances.

Alec John Jeffreys was born on January 9, 1950, in Oxford, England. He was very interested in science as a child and was given a number of pieces of scientific apparatus with which to experiment while he was still in grammar school. He enrolled at Merton College, Oxford University, from which he received his B.A. in biochemistry in 1972 and his Ph.D. in genetics in 1975. After completing his postdoctoral studies at the University of Amsterdam, Jeffreys accepted an appointment in the Department of Genetics at Leicester University, where he has remained ever since. Jeffreys sometimes pinpoints the exact moment at which he understood the potential value of RFLP technology: 9:05 a.m. on September 10, 1984. He was examining an X-ray "fingerprint" of a lab technician when he realized that such fingerprints can be used to precisely identify a person's genetic identity. He then derived the details by which such a procedure could be conducted and published the results in a paper entitled "Hypervariable 'Minisatellite' Regions in Human DNA," which appeared in the journal *Nature* on March 7, 1985.

Jeffreys has continued his research on a number of topics in genetics, including eukaryotic introns, the evolution of gene families, and the process of genetic recombination. He has received a number of honors, including election to the U.S. National Academy of Sciences, the Colworth Medal for Biochemistry of the Biochemical Society, and the Gold Medal for Zoology of the Linnean Society of London. He was knighted by Queen Elizabeth II in 1994.

John D. Kemp (b. 1940)

Kemp and Timothy Hall headed a research team that, in 1983, found a way to insert a gene from one plant, the common bean (*Phaseolus vulgaris*) into a second plant, the sunflower (*Helianthus annuus*), after which the gene was expressed as an apparently normal part of the sunflower's genome. The experiment is of significance, because it was the first time that the transfer of genes between two plants as the result of human intervention had been demonstrated, a procedure that provides the basis for much of the genetic engineering of plants conducted today.

John Daniel Kemp was born in Minneapolis, Minnesota, on January 20, 1940. He attended the University of California at Los Angeles (UCLA), from which he received his B.S. in chemistry in 1962 and his Ph.D. in biochemistry in 1965. He then remained at UCLA for postdoctoral studies from 1965 to 1968. Upon completing his postdoctoral studies, Kemp accepted an appointment as assistant professor at the University of Wisconsin, where he eventually became professor of plant pathology in 1977. During his tenure at Wisconsin, Kemp was concurrently a research chemist for the Agricultural Research Service of the USDA and vice president, head of microbiology, acting director, and associate director at the Agrigenetics Corporation in Boulder, Colorado.

In 1985, Kemp left Wisconsin to become professor of plant pathology and director of the Plant Genetic Engineering Laboratory at New Mexico State University in Las Cruces. He retired from New Mexico State in 2002 and is currently professor emeritus at the university.

Phoebus Levene (1869–1940)

Levene was a Russian-trained chemist who discovered the presence of two different types of sugar in nucleic acids, ribose and deoxyribose, from which the two compounds have been given their modern names of ribonucleic acid (RNA) and deoxyribonucleic acid (DNA), respectively.

Phoebus Aaron Theodor Levene was born as Fishel Aaronovich Levin in Sagor in Russian Lithuania, on February 25, 1869, to a family of Jewish background. His somewhat unusual first name evolved because of changes made as his family moved first to St. Petersburg, Russia (Fyodor) and then, later, to the United States (Phoebus). Levene studied medicine at the St. Petersburg Military Medical Academy, in itself an extraordinary fact since very few Jewish men were admitted to Russian institutions of higher learning. His family finally fled Russia because of rampant antisemitism there, but Levene returned long enough to complete his M.D. degree at St. Petersburg in 1891. He then returned once more to the United States.

While practicing medicine in New York City, Levene enrolled at Columbia University, where he began to take courses in biochemistry. He eventually decided to abandon medicine and pursue a research career in biochemistry. In preparation for that career, he returned to Europe, where he studied with the great German chemists Emil Fischer and Karl Kossel, from whom he developed a special interest in nucleic acids. He then returned to the United States where he took a position at the Rockefeller Institute of Medical Research (now Rockefeller University), where he spent the rest of his academic life. He was appointed head of the biochemical laboratory at Rockefeller in 1905, and it was there that he conducted the research that led to his discovery of RNA (1909) and DNA (1929). In addition, he found that the basic unit of all nucleic acids is a three-part unit, which he called a *nucleotide*, consisting of a phosphate group, a sugar (ribose or deoxyribose), and a nitrogen base. Levene died in New York City on September 6, 1940, more than a decade before the complete significance of his discoveries had become apparent.

Gregor Mendel (1822–1884)

Mendel is often called the Father of Genetics because of his research on the transmission of physical characteristics in pea plants. Based on his experiments, Mendel elucidated a set of laws, now called Mendel's Laws of Genetic Inheritance. The first law is called the Law of Segregation, namely that each parent contributes one and only one allele to the formation of an egg or sperm. The second law, the Law of Independent Assortment, which says that allele pairs in a parent separate from each other independently during reproduction, so that physical traits are also transmitted independently.

Mendel was born Johann Mendel in Heinzendorf bei Odrau, Silesia, then part of the Austrian Empire and now the city of Hynčice, Czech Republic. He worked on the family farm, where one of his chores was tending fruit trees for the lord of the local manor. Mendel attended the Philosophical Institute in Olomouc from 1840 to 1843 before entering the Augustinian Abbey of St. Thomas in Brno (Brünn). He took the name of Gregor upon becoming a monk. Mendel was ordained as a priest in 1847 and, four years later, sent to the University of Vienna to study mathematics and science in preparation for a career as a teacher in local Catholic schools. After failing his examinations at Vienna three times, he was finally assigned as a science teacher at a lower school, the Brno Realschule in 1854.

In 1857, Mendel began his historic studies of pea plants in his garden at the Brno monastery, research that continued for eight years. Over time, he grew more than 10,000 pea plants and made careful notes of the pattern of inheritance among them. His genius was to combine careful experimental skills with a skill for mathematical analysis of the results of his work. Mendel reported the results of his research at two meetings of the Natural History Society of Brno on February 8 and March 8, 1865, and then published those results in a paper, *Versuche über Pflanzen-Hybriden* ("Experiments on Plant Hybridization"), in the *Proceedings of the Natural History Society of Brno for the year 1865* (published in 1866). The scientific community paid almost no attention to Mendel's research, and his paper was cited only three times in the next 35 years. His work remained essentially unknown until it was discovered in 1900 almost simultaneously by three different researchers, Dutch botanist Hugo De Vries, Austrian botanist Erich von Tschermak, and German botanist Karl Correns.

Mendel's scientific work essentially ended in 1868 when he was elected abbot at St. Thomas. Administrative responsibilities associated with his new position, aggravated by ongoing disputes with civil authorities over taxation issues, left him little time for non-monastic issues. He died in Brno on January 6, 1884 from chronic nephritis.

Johannes Friedrich Miescher (1844–1895)

Miescher was a biochemist best known for his discovery of the first nucleic-acid-like substance, which he named *nuclein*, in 1869.

Usually known by his middle name, Friedrich Miescher was born on August 13, 1844, in Basel, Switzerland. Both his father and his uncle had held the chair of anatomy at the University of Basel, and his father encouraged Friedrich to pursue a medical degree. Friedrich decided, however, that his partial deafness, resulting from a case of typhus, might preclude a career in medicine. When he earned his M.D. from Basel in 1868, then, he decided to focus on biochemical research rather than medical practice. After graduation, he took a position in the laboratory of the eminent German biochemist Ernst Felix Hoppe-Seyler at the University of Tübingen, where he began research on the chemical composition of the cellular nucleus. In his studies, Miescher discovered a new chemical compound of unusual properties. It was similar in some way to proteins, with molecules of very large size, but it was rich in phosphorus and lacking in sulfur. Proteins never contain phosphorus, but always contain some sulfur. Hoppe-Seyler was so surprised by Miescher's results that he would not allow his student to publish until he, Hoppe-Seyler, had repeated and confirmed those results. Miescher called the substance he discovered *nuclein*, a material we now know as a complex of pure deoxyribonucleic acid (DNA) and a variety of supportive proteins typically found with DNA. In later research, Miescher was able to prepare pure DNA, although he was never able to determine its precise biochemical function.

Miescher died in Davos, Switzerland, on August 16, 1895, at the age of 51 from tuberculosis.

Kary Mullis (b. 1944)

Mullis is the inventor of a technology called the polymerase chain reaction (PCR) for the amplification of DNA segments, an accomplishment for which he was awarded the 1993 Nobel Prize in Chemistry. The technology is now a key procedure in many fields of DNA technology, because it allows the analysis of very small amounts of DNA after a series of steps that makes them large enough to study.

Kary Mullis was born in Lenoir, North Carolina, on December 24, 1944. He attended the Georgia Institute of Technology, from which he received his B.S. in 1966, and the University of California at Berkeley, where he earned his Ph.D. in biochemistry in 1973. He then completed two postdoctoral programs in pediatric cardiology at the University of Kansas Medical School and in pharmaceutical chemistry at the University of California at San Francisco. In 1979, Mullis took a position with the Cetus Corporation in Emeryville, California, one of the first companies formed to develop industrial, pharmaceutical, and other applications of new DNA technology. During his seven years at Cetus, Mullis worked on and developed the PCR technology for which he is best known. After leaving Cetus in 1986, Mullis took a job as director of molecular biology at the Xytronyx corporation in San Diego, a post he held for two years. He then began consulting on nucleic acid chemistry for a number of corporations, including Abbott Labs, Angenics, Cytometrics, Eastman Kodak, Milligen/Biosearch, and Specialty Laboratories. Mullis is also known for a somewhat free-spirited view of life and some unconventional views on scientific issues of social significance, such as the causes of HIV/AIDS disease and global climate change. He has written a popular autobiography, *Dancing Naked in the Mine Field*.

Marshall Nirenberg (b. 1927)

Nirenberg was awarded a share of the 1968 Nobel Prize in Physiology or Medicine for his discovery of the first codon, a set of three bases that codes for a specific amino acid. Earlier research had showed that the nitrogen bases on a DNA molecule acting as a triplet direct the synthesis of specific amino acids. That is, the combination CGA (cytosine-guanine-adenine) somehow "tells" a cell to make a specific amino acid. The problem was that no one yet knew which codons coded for which amino acids. Working with Heinrich J. Matthaei at the NIH, Nirenberg obtained the answer to that question. The Nirenberg-Matthaei experiment was conceptually relatively simple. They prepared RNA molecules consisting only of the nitrogen base uracil (UUU). They then inserted those molecules into a sample of *E. coli* bacteria which was treated to destroy their natural DNA, ensuring that the only compounds synthesized were those coded for by the UUU triplet. Upon analysis, Nirenberg and Matthaei found that the only compound produced by the bacterial cells was the amino acid phenylalanine. It was evident, then, that the triplet UUU codes for the amino acid phenylalanine. The first element in the genetic code had been determined.

Marshall Warren Nirenberg was born in New York City on April 10, 1927. When he developed rheumatic fever, the family moved to Florida, where Marshall soon developed an intense interest in the natural world. He began to collect an extensive set of notes and diaries about his own research on living organisms. In 1945, Nirenberg enrolled at the University of Florida at Gainesville, from which he received his B.S. in chemistry and zoology in 1948 and his M.S. in zoology in 1952. Nirenberg then continued his studies at the University of Michigan in Ann Arbor, which granted him his Ph.D. in biological chemistry in 1957. He then completed two postdoctoral programs at the National Institute of Arthritis, Metabolism, and Digestive Diseases (NIAMDD). It was at NIAMDD that Nirenberg carried out his work on the genetic code with Matthaei.

In 1962, Nirenberg was offered the position of chief of the section of biochemical genetics at the National Heart Institute (now the National Heart, Lung, and Blood Institute; NHLBI) of the NIH. He has remained with the NHLBI ever since. In addition to the Nobel Prize, Nirenberg has been awarded the National Medal of Science, the National Medal of Honor, and the Molecular Biology Award of the National Academy of Sciences. In 2001 he was elected to the American Philosophical Society.

Ingo Potrykus (b. 1933)

In collaboration with Peter Beyer, Potrykus invented golden rice, a genetically engineered agricultural product capable of delivering increased amounts of vitamin A to people who eat it. The product potentially has very significant benefits since as many as 250,000 to 500,000 children around the world suffer from blindness and death because of suboptimal levels of vitamin A in their daily diet.

Ingo Potrykus was born in Hirschberg, Germany, on December 5, 1933. He attended the universities of Cologne and Erlangen, where he studied botany, biochemistry, zoology, genetics, philosophy, and physical education. He received his Ph.D. in plant genetics from the Max-Planck-Institute for Plant Breeding Research in Cologne in 1968. He was then offered a position at the Institute of Plant Physiology in Stuttgart-Hohenheim, where he remained from 1970 to 1974. Potrykus then took a position as research group leader at the Max-Planck-Institute for Genetics at Ladenburg-Heidelberg from 1974 to 1976 before assuming a similar position at the Friedrich Miescher Institute in Basel, Switzerland. In 1986, he left the Friedrich Miescher Institute to become professor of plant sciences at the Eidgenössische Technische Hochschule Zürich (Swiss Federal Institute of Technology; ETH), where he remained until his retirement in 1999.

Throughout his career, Potrykus has been interested in finding ways of using genetic engineering to increase food production for the world's growing population. He was especially interested in solving the problem of adding vitamin A to the rice grains on which so much of the world's population depends for its primary food. Rice plants do synthesize the precursors of vitamin A in their vegetative tissues, but not in rice grains themselves. Potrykus and his colleagues found that the addition of genes for two additional enzymes to the rice plant, phytoene synthase and phytoene desaturase, made possible the production of vitamin A in rice grains as well as in the plant's tissues.

Solving the scientific problem of producing golden rice was only the beginning for Potrykus and his followers. Since laboratory success was achieved, the Golden Rice project has been attempting to complete other requirements needed before the product can actually be used in the field, such as gaining necessary financial support and licenses and conducting field and nutrition tests. As of 2009, the product is still not generally available for the public, although scientific progress continues. In 2004, a new form of golden rice was announced that provides 23 times the amount of vitamin A as that available from Potrykus's original engineered plant.

Robert A. Swanson (1947–1999)

Swanson was a venture capitalist who founded the biotechnology firm Genentech in 1976 with Herbert W. Boyer. He was affiliated with the company for almost the rest of his life, retiring in December 1996.

Robert Arthur Swanson was born on November 29, 1947, in Brooklyn, New York. He attended the Massachusetts Institute of Technology (MIT), from which he received his S.B. in chemistry and S.M. in business management, both in 1970. He then took a job with Citicorp Venture Capital Limited, where he remained until 1974. He then joined the venture capital firm of Kleiner, Perkins, Caufield & Byers (KPDB) in Menlo Park, California. A year later, aided by $100,000 in seed money from KPDB, Swanson and Boyer founded Genentech for the purpose of developing new recombinant DNA products and developing them for the marketplace. He served as chief executive officer at Genentech from 1976 to 1990 and as chairman of the board from 1990 to 1996. Swanson then left Genentech to continue venture capital activities on his own, eventually becoming involved in the formation of a new biotechnology company, Tularik, Inc. In 1998, Swanson fell ill with brain cancer, a disease from which he died on December 6, 1999, at his home in Hillsborough, California.

Hamilton O. Smith (b. 1931)

Smith was awarded a share of the 1978 Nobel Prize for Physiology or Medicine for his research on restriction enzymes. Restriction enzymes are chemical compounds that recognize specific nucleotide sequences in a nucleic acid and then cleave the molecule at those sites. More than 3,000 restriction enzymes are currently known and about 600 are commercially available. Restriction enzymes are essential in most DNA technologies, because they allow a researcher to cut DNA and RNA molecules in precise locations, permitting the insertion of DNA or RNA segments from other sources in the production of transgenic organisms. In his research, Smith and his colleagues identified the first restriction enzyme, found in the bacterium *Haemophilus influenzae*, which they called endonuclease R, but which is now known as HindII. Shortly thereafter, they also discovered the DNA sequence that the enzyme recognizes and cuts:

GT(T/C)(A/G)AC
CA(A/G)(T/C)TG

Hamilton Othanel Smith was born in New York City on August 23, 1931. In 1937, his father joined the faculty of the Department of Education at the University of Illinois (UI), and Hamilton became a student at the University Laboratory School at UI. After he graduated from high school, Smith enrolled at the UI, planning to major in mathematics, for which, he later wrote, "I had a flair but no deep talent." In 1950, he transferred to the University of California at Berkeley (UCB), where, although he continued his studies in mathematics, he discovered that his real passion was biology and biochemistry. After receiving his degree from UCB in 1952, Smith applied to the Johns Hopkins University Medical School and was accepted. He was awarded his M.D. there in 1956.

After completing his internship at Barnes Hospital in St. Louis, his residency at University Hospital at the University of Michigan, and a tour of duty as researcher in human genetics at Michigan, Smith accepted an appointment as assistant professor of microbiology at Johns Hopkins University. He remained at Johns Hopkins until his retirement in 1998, when he was named American Cancer Society Distinguished Research Professor Emeritus of Molecular Biology and Genetics. In 1994, Smith collaborated with J. Craig Venter at The Institute for Genomic Research to sequence the complete genome of the *H. influenzae* bacterium. In July 1998, Smith joined a research team at Celera Genomics , where he worked on sequencing the human genome and the genome of the common fruit fly. In November 2002, he left Celera to join the J. Craig Venter Institute, where he is leader of the synthetic biology and bioenergy research groups.

J. Craig Venter (b. 1946)

While a researcher at the NIH in the 1980s, Venter developed a method for mapping the human genome using chemical units known as expressed sequence tags (ESTs). He founded Celera Genomics in 1998 as a vehicle for using ESTs to elucidate the human genome, a program that eventually ran concurrently for many years with the Human Genome Project.

John Craig Venter was born in Salt Lake City, Utah, on October 14, 1946. He attended the College of San Mateo (a community college) and the University of California at San Diego, from which he received his B.S. in biochemistry and his Ph.D. in physiology and pharmacology in 1972 and 1975, respectively. He then joined the State University of New York at Buffalo, where he held a variety of posts, also serving concurrently as a cancer researcher at the Roswell Park Memorial Institute. In 1984, he left Buffalo to join the NIH, where he discovered ESTs and their potential for mapping a genome.

In 1992, Venter left the NIH to found his own research institution, The Institute for Genomic Research (TIGR), with the goal of using his own method of gene sequencing (EST) to map the genome of organisms. By 1995, researchers at TIGR had decoded the genome of the first free-living organism, the bacterium *Haemophilus influenzae*. Three years later, Venter founded Celera Genomics for the purpose of sequencing the human genome. He hoped that his approach to mapping would allow him to elucidate the human genome before the federal Human Genome Project could achieve the same goal. Eventually, researchers at Celera and the Human Genome Project began to collaborate in their work and in 2001, they published their results jointly in papers published in the journals *Science* and *Nature*. Since mapping of the human genome has been completed, Celera researchers have also sequenced a number of other genomes, including those of the fruit fly, mouse, rat, cat, dog, African elephant, and common chimpanzee. In June 2005, Venter cofounded Synthetic Genomics, a firm dedicated to the use of genetically modified microorganisms for the production of clean fuels and biochemicals.

James Watson (b. 1928)

In 1953, Watson and British chemist Francis Crick announced their model for the structure of a DNA molecule. They suggested that the molecule was a double helix—two strands of nucleotides wrapped around each other—consisting of long chains made of alternating phosphate and sugar (deoxyribose) groups, joined to each other by pairs of nitrogen bases. Some historians have called this discovery the most important accomplishment in the history of biology. Since 1953, Watson has devoted the greatest part of his life to administrative assignments, primarily at the Cold Spring Harbor Laboratory, on Long Island, New York. From 1989 to 1992 he was head of the Human Genome Project.

James Dewey Watson was born on April 6, 1928, in Chicago, Illinois. He demonstrated extraordinary intellectual skills at an early age and appeared frequently on a popular radio show of the time called *Quiz Kids*. He entered the University of Chicago at the age of 15, where he planned to major in his longtime passion of ornithology. After reading Erwin Schrödinger's popular and influential book, *What Is Life?*, however, he changed his major to genetics. He received his B.S. in zoology from Chicago in 1947 and then continued his studies at Indiana University under Salvador Luria, earning his Ph.D. in 1950 at the age of 22.

After completing a postdoctoral year of study at the University of Copenhagen, Watson accepted a second postdoctoral appointment at the world-famous Cavendish Laboratory at the University of Cambridge. There he met Francis Crick and discovered that the two men shared a passion for unraveling the structure of the DNA molecule. The complex series of events that led to their success has been recounted in Watson's book, *The Double Helix*.

In 1953, Watson returned to the United States and took a post as senior research fellow at the California Institute of Technology. He then spent another year at the Cavendish before accepting an appointment at Harvard University, where he spent the rest of his academic career. He received the 1962 Nobel Prize in Physiology or Medicine, along with Crick, for their discovery of the structure of DNA. He has also received more than two dozen honorary degrees, and a large number of other awards and prizes.

Ray Wu (1928–2008)

In the 1970s, Wu developed methods for determining the nucleotide sequence in DNA molecules, which have made possible the sequencing of an organism's genome. With a number of modifications, Wu's procedures still form the basis of genome sequencing today. Wu also developed a number of fundamental techniques for the production of transgenic plants. He has been especially interested in developing engineered forms of the rice plant that are resistant to pests, drought, salt water, and temperature extremes. In the mid-1990s, for example, Wu developed a method for transferring from the potato plant the gene for an enzyme, protease inhibitor II, into a rice plant. That gene promotes the production of a protein that interferes with the metabolism of the pink stem borer, a major pest of rice plants. The bioengineered rice plant thus produced was, therefore, resistant to attack by the borer. He later developed methods for engineering rice plants not only to make them tolerant to salt water, drought, and temperature extremes, but also to increase their yield.

Ray Jui Wu was born in Peking, China, on August 14, 1928. He moved to the United States in 1948 to receive his college education, earning his B.S. in chemistry at the University of Alabama (where his father was a biochemist) in 1950 and his Ph.D. in biochemistry at the University of Pennsylvania in 1955. From 1955 to 1966 he was Damon Runyon research fellow, assistant professor, associate professor, and associate member at the Public Health Research Institute in New York City. He then accepted an appointment as associate professor of biochemistry and molecular biology at Cornell University, where he was promoted to full professor in 1972. In 2004 he was named Liberty Hyde Bailey Professor of Molecular Biology and Genetics, a post he held until his death of heart failure on February 10, 2008, in Ithaca, New York.

Wu became a U.S. citizen in 1961 but continued to maintain close ties with scientific communities in China and Taiwan. He was an honorary professor at the Peking Medical College, Fudan University, Wuhan University, Zhonmgshan University, Beijing Agricultural University, and Fujian Agricultural University. In 1982, Wu organized the China-United States Biochemistry and

Molecular Biology Examination and Application program, a program that brought gifted Chinese students to the United States for study. That effort later grew into a larger effort to promote the cooperation and collaboration of Chinese and American life scientists, an effort called the Ray Wu Society (now the Chinese Biological Investigator Society).

6

Data and Documents

This chapter contains a number of important documents dealing with DNA technology and its applications in a variety of fields, along with some data and statistics dealing with those issues. At this point in history, national, international, and state governments have passed relatively few laws or adopted few regulations dealing with the use of DNA technology. More common has been the promulgation of policies on DNA technology in general or on some aspect of the technology, such as genetic testing or human gene therapy. The documents and data are arranged according to type: laws and regulations from both the United States and other parts of the world, declarations and position statements, and reports from governmental agencies on one or another form of DNA technology. (An asterisk and ellipses [* . . .] indicate omitted citations.)

Laws and Regulations

Coordinated Framework for Regulation of Biotechnology (1986)

In 1986, all agencies of the federal government with responsibility for regulating any aspect of genetically engineered organisms combined to develop a common framework for their efforts. That document, a 123-page paper, continues to describe the distribution of tasks in the regulation of biotechnology for each of the many agencies involved in that process. Two charts included in the report summarize the assignment of responsibilities to agencies.

TABLE 6.1
Coordinated Framework for Regulation of Biotechnology

CHART I. —COORDINATED FRAMEWORK—APPROVAL OF COMMERCIAL BIOTECHNOLOGY PRODUCTS	
Subject	**Responsible agency(ies)**
Foods/Food Additives	FDA* FSIS[1]
Human Drugs, Medical Devices and Biologics	FDA
Animal Drugs	FDA
Animal Biologics	APHIS
Other Contained Uses	EPA
Plants and Animals	APHIS* FSIS[1] FDA[2]
Pesticide Microorganisms Released in the Environment All	EPA* APHIS[3]
Other Uses (Microorganisms)	
Intergeneric Combination	EPA* APHIS[3]
Intrageneric Combination Pathogenic Source Organism	
Agricultural Use	APHIS
Non-Agricultural Use	EPA*[4] APHIS
No Pathogenic Source Organisms	EPA Report
Nonengineered Pathogens:	
Agricultural Use	APHIS
Non-Agricultural Use	EPA[4] APHIS[3]
Nonengineered Nonpathogens	EPA Report

*Lead agency

[1]FSIS (Food Safety and Inspection Service) under the Assistant Secretary of Agriculture for Marketing and Inspection Services is responsible for food use.

[2]FDA (Food and Drug Administration) is involved when in relation to a food use.

[3]APHIS (Animal and Plant Health Inspection Service) is involved when the microorganism is plant pest, animal pathogen, or regulated article requiring a permit.

[4]EPA (Environmental Protection Agency) requirements will only apply to environmental release under a "significant new use rule" that EPA intends to propose.

CHART II. —COORDINATED FRAMEWORK—BIOTECHNOLOGY RESEARCH JURISDICTION

Subject	Responsible agency(ies)
Contained Research, No Release in Environment	
Federally funded	Funding agency[1]
Non-federally funded	NIH or S&E voluntary review: APHIS[2]
Foods/Food Additives, Human Drugs, Medical Devices, Biologics, Animal Drugs	
Federally funded	FDA* NIH guidelines & review
Non-federally funded	FDA* NIH voluntary review
Plants, Animals, and Animal Biologics	
Federally funded	Funding agency[1] APHIS[2]
Non-federally funded	APHIS* S&E voluntary review
Pesticide Microorganisms Genetically Engineered	
Intergenetic	EPA* APHIS[2]S&E voluntary review
Pathogenic Intergenetic	EPA* APHIS[2]S&E voluntary review
Intrageneric Nonpathogen Nonengineered	
Nonindigenous Pathogens	EPA* APHIS
Indigenous Pathogens	EPA*[3] APHIS
Nonindigenous Nonpathogen	EPA*
Other Uses (Microorganisms) Released in the Environment Genetically Engineered Intergenetic Organisms	
Federally funded	Funding agency*[1] APHIS[2] EPA[4]
Commercially funded	EPA APHIS S&E voluntary review
Intrageneric Organisms Pathogenic Source Organism	
Federally funded	Funding agency*[1] APHIS[2] EPA[4]

TABLE 6.1 continued

CHART II. —COORDINATED FRAMEWORK—BIOTECHNOLOGY RESEARCH JURISDICTION	
Subject	**Responsible agency(ies)**
Commercially funded	APHIS[2] EPA (*if nonagriculture use)
Intrageneric Combination	
No Pathogenic Source Organisms	EPA Report
Nonengineered	EPA Report* APHIS[2]

*Lead agency

[1]Review and approval of research protocols conducted by NIH, S&E, or NSF.

[2]APHIS issues permits for the importation and domestic shipment of certain plants and animals, plant pests and animal pathogens, and for the shipment or release in the environment of regulated articles.

[3]EPA jurisdiction for research on a plot greater than 10 acres.

[4]EPA reviews federally funded environmental research only when it is for commercial purposes.

Source: Office of Science and Technology Policy. *Coordinated Framework for Regulation of Biotechnology.* 51 FR 23302, June 26, 1986. Available online. URL: http://usbiotechreg.nbii.gov/Coordinated_Framework_1986_Federal_Register.html. Accessed on March 21, 2009.

Statutory Requirements for DNA Testing (1994)

In 1994, the U.S. Congress passed amendments to the Omnibus Crime Control and Safe Streets Act of 1968 containing provisions pertaining to DNA testing. Those provisions then became incorporated into the U.S. Code, Title 42, Chapter 136, Subchapter IX, Part A, as Sections 14131–14133. Those section are excerpted here in their updated form.

§ 14131. Quality assurance and proficiency testing standards

(a) Publication of quality assurance and proficiency testing standards

  (1)

    (A) Not later than 180 days after September 13, 1994, the Director of the Federal Bureau of Investigation shall appoint an advisory board on DNA quality assurance methods from among nominations proposed by the head of the National Academy of Sciences and professional societies of crime laboratory officials.

(B) The advisory board shall include as members scientists from State, local, and private forensic laboratories, molecular geneticists and population geneticists not affiliated with a forensic laboratory, and a representative from the National Institute of Standards and Technology.

(C) The advisory board shall develop, and if appropriate, periodically revise, recommended standards for quality assurance, including standards for testing the proficiency of forensic laboratories, and forensic analysts, in conducting analyses of DNA.

(2) The Director of the Federal Bureau of Investigation, after taking into consideration such recommended standards, shall issue (and revise from time to time) standards for quality assurance, including standards for testing the proficiency of forensic laboratories, and forensic analysts, in conducting analyses of DNA.

(3) The standards described in paragraphs (1) and (2) shall specify criteria for quality assurance and proficiency tests to be applied to the various types of DNA analyses used by forensic laboratories. The standards shall also include a system for grading proficiency testing performance to determine whether a laboratory is performing acceptably.

(4) Until such time as the advisory board has made recommendations to the Director of the Federal Bureau of Investigation and the Director has acted upon those recommendations, the quality assurance guidelines adopted by the technical working group on DNA analysis methods shall be deemed the Director's standards for purposes of this section.

* ... [*Section b deals with the administration of the board.*]

(c) Proficiency testing program

(1) Not later than 1 year after the effective date of this Act,[1] the Director of the National Institute of Justice shall certify to the Committees on the Judiciary of the House and Senate that—

(A) the Institute has entered into a contract with, or made a grant to, an appropriate entity for establishing, or has taken other appropriate action to ensure that there is established, not later than 2 years after September 13, 1994, a blind external proficiency testing program for DNA analyses, which shall be available to public and private laboratories performing forensic DNA analyses;

(B) a blind external proficiency testing program for DNA analyses is already readily available to public and private laboratories performing forensic DNA analyses; or

(C) it is not feasible to have blind external testing for DNA forensic analyses.

(2) As used in this subsection, the term "blind external proficiency test" means a test that is presented to a forensic laboratory through a second agency and appears to the analysts to involve routine evidence.

(3) Notwithstanding any other provision of law, the Attorney General shall make available to the Director of the National Institute of Justice during the first fiscal year in which funds are distributed under this subtitle up to $ 250,000 from the funds available under part X of Title I of the Omnibus Crime Control and Safe Streets Act of 1968 [42 U.S.C. 3796kk et seq.] to carry out this subsection.

§ 14132. Index to facilitate law enforcement exchange of DNA identification information

(a) Establishment of index

The Director of the Federal Bureau of Investigation may establish an index of—

(1) DNA identification records of—

(A) persons convicted of crimes;

(B) persons who have been charged in an indictment or information with a crime; and

(C) other persons whose DNA samples are collected under applicable legal authorities, provided that DNA samples that are voluntarily submitted solely for elimination purposes shall not be included in the National DNA Index System;

(2) analyses of DNA samples recovered from crime scenes;

(3) analyses of DNA samples recovered from unidentified human remains; and

(4) analyses of DNA samples voluntarily contributed from relatives of missing persons.

(b) Information

The index described in subsection (a) of this section shall include only information on DNA identification records and DNA analyses that are—

(1) based on analyses performed by or on behalf of a criminal justice agency (or the Secretary of Defense in accordance with section 1565 of title 10) in accordance with publicly available standards that satisfy or exceed the guidelines for a quality assurance program for DNA analysis, issued by the Director of the Federal Bureau of Investigation under section 14131 of this title;

(2) prepared by laboratories that—

(A) not later than 2 years after October 30, 2004, have been accredited by a nonprofit professional association of persons actively involved in forensic science that is nationally recognized within the forensic science community; and

(B) undergo external audits, not less than once every 2 years, that demonstrate compliance with standards established by the Director of the Federal Bureau of Investigation; and

(3) maintained by Federal, State, and local criminal justice agencies (or the Secretary of Defense in accordance with section 1565 of title 10) pursuant to rules that allow disclosure of stored DNA samples and DNA analyses only—

(A) to criminal justice agencies for law enforcement identification purposes;

(B) in judicial proceedings, if otherwise admissible pursuant to applicable statutes or rules;

(C) for criminal defense purposes, to a defendant, who shall have access to samples and analyses performed in connection with the case in which such defendant is charged; or

(D) if personally identifiable information is removed, for a population statistics database, for identification research and protocol development purposes, or for quality control purposes.

(c) Failure to comply

Access to the index established by this section is subject to cancellation if the quality control and privacy requirements described in subsection (b) of this section are not met.

(d) Expungement of records

(1) By Director

(A) The Director of the Federal Bureau of Investigation shall promptly expunge from the index described in subsection (a) of this section the DNA analysis of a person included in the index—

(i) on the basis of conviction for a qualifying Federal offense or a qualifying District of Columbia offense (as determined under sections 14135a and 14135b of this title, respectively), if the Director receives, for each conviction of the person of a qualifying offense, a certified copy of a final court order establishing that such conviction has been overturned; or

(ii) on the basis of an arrest under the authority of the United States, if the Attorney General receives, for each charge against the person on the basis of which the analysis was or could have been included in the index, a certified copy of a final court order establishing that such charge has been dismissed or has resulted in an acquittal or that no charge was filed within the applicable time period.

(B) For purposes of subparagraph (A), the term "qualifying offense" means any of the following offenses:

(i) A qualifying Federal offense, as determined under section 14135a of this title.

(ii) A qualifying District of Columbia offense, as determined under section 14135b of this title.

(iii) A qualifying military offense, as determined under section 1565 of title 10.

(C) For purposes of subparagraph (A), a court order is not "final" if time remains for an appeal or application for discretionary review with respect to the order.

(2) By States

(A) As a condition of access to the index described in subsection (a) of this section, a State shall promptly expunge from that index the DNA analysis of a person included in the index by that State if—

(i) the responsible agency or official of that State receives, for each conviction of the person of an offense on the basis of which that analysis was or could have been included in the index, a certified copy of a final court order establishing that such conviction has been overturned; or

(ii) the person has not been convicted of an offense on the basis of which that analysis was or could have been included in the index, and the responsible agency or official of that State receives, for each charge against the person on the basis of which the analysis was or could have been included in the index, a certified copy of a final court order establishing that such charge has been dismissed or has resulted in an acquittal or that no charge was filed within the applicable time period.

(B) For purposes of subparagraph (A), a court order is not "final" if time remains for an appeal or application for discretionary review with respect to the order.

§ 14133. Federal Bureau of Investigation

(a) Proficiency testing requirements

(1) Generally

(A) Personnel at the Federal Bureau of Investigation who perform DNA analyses shall undergo semiannual external proficiency testing by a DNA proficiency testing program meeting the standards issued under section 14131 of this title.

(B) Within 1 year after September 13, 1994, the Director of the Federal Bureau of Investigation shall arrange for periodic blind external tests to determine the proficiency of DNA analysis performed at the Federal Bureau of Investigation laboratory.

(C) In this paragraph, "blind external test" means a test that is presented to the laboratory through a second agency and appears to the analysts to involve routine evidence.

(2) Report

For 5 years after September 13, 1994, the Director of the Federal Bureau of Investigation shall submit to the Committees on the Judiciary of the House and Senate an annual report on the results of each of the tests described in paragraph (1).

(b) Privacy protection standards

(1) Generally

Except as provided in paragraph (2), the results of DNA tests performed for a Federal law enforcement agency for law enforcement purposes may be disclosed only—

(A) to criminal justice agencies for law enforcement identification purposes;

(B) in judicial proceedings, if otherwise admissible pursuant to applicable statues [sic] or rules; and

(C) for criminal defense purposes, to a defendant, who shall have access to samples and analyses performed in connection with the case in which such defendant is charged.

(2) Exception

If personally identifiable information is removed, test results may be disclosed for a population statistics database, for identification research and protocol development purposes, or for quality control purposes.

(c) Criminal penalty

(1) A person who—

(A) by virtue of employment or official position, has possession of, or access to, individually identifiable DNA information indexed in a database created or maintained by any Federal law enforcement agency; and

(B) knowingly discloses such information in any manner to any person or agency not authorized to receive it,

shall be fined not more than $ 100,000.

(2) A person who, without authorization, knowingly obtains DNA samples or individually identifiable DNA information indexed in a database created or maintained by any Federal law enforcement agency shall be fined not more than $ 250,000, or imprisoned for a period of not more than one year, or both.

Source: U.S. Code, Title 42, Chapter 136, Subchapter IX, Part A, §14131-14133. Available online. URL: http://www.law.cornell.edu/uscode/html/uscode42/usc_sec_42_00014131——000-.html-http://www.law.cornell.edu/uscode/html/uscode42/usc_sec_42_00014133——000-.html.

Executive Order 13145 (2000)

Federal policy on the use of information obtained from genetic testing is enshrined in two documents, a Presidential executive order signed by President Bill Clinton on February 8, 2000, and the Genetic Information Nondiscrimination Act of 2008 (see below). Clinton's order lays out federal policy on the use of genetic information, describes the type of information covered in the order, and lists certain exceptions to the requirements posed in the order. The most relevant parts of the order are as follows:

Section 1. Nondiscrimination in Federal Employment on the Basis of Protected Genetic Information.

1-101. It is the policy of the Government of the United States to provide equal employment opportunity in Federal employment for all qualified persons and to prohibit discrimination against employees based on protected genetic information, or information about a request for or the receipt of genetic services. This policy of equal opportunity applies to every aspect of Federal employment.

1-102. The head of each Executive department and agency shall extend the policy set forth in section 1-101 to all its employees covered by section 717 of Title VII of the Civil Rights Act of 1964, as amended (42 U.S.C. 2000e-16).

1-103. Executive departments and agencies shall carry out the provisions of this order to the extent permitted by law and consistent with their statutory and regulatory authorities, and their enforcement mechanisms. The Equal Employment Opportunity Commission shall be responsible for coordinating the policy of the Government of the United States to prohibit discrimination against employees in Federal employment based on protected genetic information, or information about a request for or the receipt of genetic services.

* . . . [*Section 2 list definitions of terms used in the order and provides specific nondiscrimination responsibilities for departments and agencies. Section 3 lists exceptions allowed to the execution of this order. Section 4 deals with miscellaneous issues.*]

Source: "Executive Order: February 8, 2000." Available online. URL: http://frwebgate.access.gpo.gov/cgi-bin/getdoc.cgi?dbname=2000 _register&docid=fr10fe00-165.pdf. Accessed on March 17, 2009.

Voluntary Labeling Indicating Whether Foods Have or Have Not Been Developed Using Bioengineering (2001)

The U.S. government currently has no laws requiring the labeling of genetically modified foods. The possibility of such legislation has, however, been considered by the U.S. Congress for some time, and bills to that effect have regularly been introduced into the Congress. Thus far, the regulation most directly related to this subject is a draft issued by the Food and Drug Administration in January 2001. Some pertinent sections of that guideline are as follows (references have been omitted):

In determining whether a food is misbranded, FDA would review label statements about the use of bioengineering to develop a food or its ingredients under sections 403(a) and 201(n) of the act [Federal Food, Drug, and Cosmetic Act]. Under section 403(a) of the act, a food is misbranded if statements on its label or in its labeling are false or misleading in any particular. Under section 201(n), both the presence and the absence of information are relevant to whether labeling is misleading. That is, labeling may be misleading if it fails to disclose facts that are material in light of representations made about a product or facts that are material with respect to the consequences that may result from use of the product. In determining whether a statement that a food is or is not genetically engineered is misleading under sections 201(n) and 403(a) of the act, the agency will take into account the entire label and labeling.

Statements about foods developed using bioengineering

FDA recognizes that some manufacturers may want to use informative statements on labels and in labeling of bioengineered foods or foods that contain ingredients produced from bioengineered foods. The following are examples of some statements that might be used. The discussion accompanying each example is intended to provide guidance as to how similar statements can be made without being misleading.

- "Genetically engineered" or "This product contains cornmeal that was produced using biotechnology."

The information that the food was bioengineered is optional and this kind of simple statement is not likely to be misleading. However, focus group data indicate that consumers would prefer label statements that disclose and explain the goal of the technology (why it was used or what it does for/to the food). Consumers also expressed some preference for the term "biotechnology" over such terms as "genetic modification" and "genetic engineering".

- "This product contains high oleic acid soybean oil from soybeans developed using biotechnology to decrease the amount of saturated fat."

This example includes both required and optional information. As discussed above in the background section, when a food differs from its traditional counterpart such that the common or usual name no longer adequately describes the new food, the name must be changed to describe the difference. Because this soybean oil contains more oleic acid than traditional soybean oil, the term "soybean oil" no longer adequately describes the nature of the food. Under section 403(i) of the act, a phrase like "high oleic acid" would be required to appear as part of the name of the food to describe its basic nature. The statement that the soybeans were developed using biotechnology is optional. So is the statement that the reason for the change in the soybeans was to reduce saturated fat.

- "These tomatoes were genetically engineered to improve texture."

In this example, the change in texture is a difference that may have to be described on the label. If the texture improvement makes a significant difference in the finished product, sections 201(n) and 403(a)(1) of the act would require disclosure of the difference for the consumer. However, the statement must not be misleading. The phrase "to improve texture" could be misleading if the texture difference is not noticeable to the consumer. For example, if a manufacturer wanted to describe a difference in a food that the consumer would not notice when purchasing or consuming the product, the manufacturer should phrase the statements so that the consumer can understand the significance of the difference. If the change in the tomatoes was intended to facilitate processing but did not make a noticeable difference in the processed consumer product, a phrase like "to improve texture for processing" rather than "to improve texture" should be used to ensure that the consumer is not misled. The statement that the tomatoes were genetically engineered is optional.

- "Some of our growers plant tomato seeds that were developed through biotechnology to increase crop yield."

The entire statement in this example is optional information. The fact that there was increased yield does not affect the characteristics of the food and is therefore not necessary on the label to adequately describe the food for the consumer. A phrase like "to increase yield" should only be included where there is substantiation that there is in fact the stated difference.

Where a benefit from a bioengineered ingredient in a multi-ingredient food is described, the statement should be worded so that it addresses the ingredient and not the food as a whole; for example, "This product contains high oleic acid soybean oil from soybeans produced through biotechnology to decrease the level of saturated fat." In addition, the amount of the bioengineered ingredient in the food may be relevant to whether the statement is misleading. This would apply especially where the bioengineered difference is a nutritional improvement. For example, it would likely be misleading to make a statement about a nutritionally improved ingredient on a food that contains only a small amount of the ingredient, such that the food's overall nutritional quality would not be significantly improved.

FDA reminds manufacturers that the optional terms that describe an ingredient of a multi-ingredient food as bioengineered should not be used in the ingredient list of the multi-ingredient food. Section 403 (i)(2) of the act requires each ingredient to be declared in the ingredient statement by its common or usual name. Thus, any terms not part of the name of the ingredient are not permitted in the ingredient statement.

In addition, 21 CFR 101.2(e) requires that the ingredient list and certain other mandatory information appear in one place without other intervening material. FDA has long interpreted any optional description of ingredients in the ingredient statement to be intervening material that violates this regulation.

Statements about foods that are not bioengineered or that do not contain ingredients produced from bioengineered foods

Terms that are frequently mentioned in discussions about labeling foods with respect to bioengineering include "GMO free" and "GM free." "GMO" is an acronym for "genetically modified organism" and "GM" means "genetically modified." Consumer focus group data indicate that consumers do not understand the acronyms "GMO" and "GM" and prefer label statements with spelled out words that mean bioengineering.

Terms like "not genetically modified" and "GMO free," that include the word "modified" are not technically accurate unless they are clearly in a context that refers to bioengineering technology. "Genetic modification" means the alteration of the genotype of a plant using any technique, new or traditional. "Modification" has a broad context that means the alteration in the composition of food that results from adding, deleting, or changing hereditary traits, irrespective of the method. Modifications may be minor, such as a single mutation that affects one gene, or major alterations of genetic material that affect many genes. Most, if not all, cultivated food crops have been genetically modified. Data indicate that consumers do not have a good understanding that essentially all food crops have been genetically modified and that bioengineering technology is only one of a number of technologies used to genetically modify crops. Thus, while it is accurate to say that a bioengineered food was "genetically modified," it likely would be inaccurate to state that a food that had not been produced using biotechnology was "not genetically modified" without clearly providing a context so that the consumer can understand that the statement applies to bioengineering.

The term "GMO free" may be misleading on most foods, because most foods do not contain organisms (seeds and foods like yogurt that contain microorganisms are exceptions). It would likely be misleading to suggest that a food that ordinarily would not contain entire "organisms" is "organism free."

There is potential for the term "free" in a claim for absence of bioengineering to be inaccurate. Consumers assume that "free" of bioengineered material means that "zero" bioengineered material is present. Because of the potential for adventitious presence of bioengineered material, it may be necessary to conclude that the accuracy of the term "free" can only be ensured when there is a definition or threshold above which the term could not be used. FDA does not have information with which to establish a threshold level of bioengineered constituents or

ingredients in foods for the statement "free of bioengineered material."
FDA recognizes that there are analytical methods capable of detecting
low levels of some bioengineered materials in some foods, but a threshold would require methods to test for a wide range of genetic changes at
very low levels in a wide variety of foods. Such test methods are not
available at this time. The agency suggests that the term "free" either
not be used in bioengineering label statements or that it be in a context
that makes clear that a zero level of bioengineered material is not
implied. However, statements that the food or its ingredients, as appropriate, was not developed using bioengineering would avoid or minimize such implications. For example,

- "We do not use ingredients that were produced using
  biotechnology;"
- "This oil is made from soybeans that were not genetically
  engineered;" or
- "Our tomato growers do not plant seeds developed
  using biotechnology."

A statement that a food was not bioengineered or does not contain
bioengineered ingredients may be misleading if it implies that the
labeled food is superior to foods that are not so labeled. FDA has concluded that the use or absence of use of bioengineering in the production
of a food or ingredient does not, in and of itself, mean that there is a
material difference in the food. Therefore, a label statement that
expresses or implies that a food is superior (e.g., safer or of higher quality) because it is not bioengineered would be misleading. The agency
will evaluate the entire label and labeling in determining whether a label
statement is in a context that implies that the food is superior.

In addition, a statement that an ingredient was not bioengineered
could be misleading if there is another ingredient in the food that was
bioengineered. The claim must not misrepresent the absence of bioengineered material. For example, on a product made largely of bioengineered corn flour and a small amount of soybean oil, a claim that the
product "does not include genetically engineered soybean oil" could
be misleading. Even if the statement is true, it is likely to be misleading
if consumers believe that the entire product or a larger portion of it than
is actually the case is free of bioengineered material. It may be necessary
to carefully qualify the statement in order to ensure that consumers
understand its significance.

Further, a statement may be misleading if it suggests that a food or
ingredient itself is not bioengineered, when there are no marketed bioengineered varieties of that category of foods or ingredients. For example, it would be misleading to state "not produced through
biotechnology" on the label of green beans, when there are no marketed

bioengineered green beans. To not be misleading, the claim should be in a context that applies to the food type instead of the individual manufacturer's product. For example, the statement "green beans are not produced using biotechnology" would not imply that this manufacturer's product is different from other green beans.

Substantiation of label statements

A manufacturer who claims that a food or its ingredients, including foods such as raw agricultural commodities, is not bioengineered should be able to substantiate that the claim is truthful and not misleading. Validated testing, if available, is the most reliable way to identify bioengineered foods or food ingredients. For many foods, however, particularly for highly processed foods such as oils, it may be difficult to differentiate by validated analytical methods between bioengineered foods and food ingredients and those obtained using traditional breeding methods. Where tests have been validated and shown to be reliable they may be used. However, if validated test methods are not available or reliable because of the way foods are produced or processed, it may be important to document the source of such foods differently. Also, special handling may be appropriate to maintain segregation of bioengineered and nonbioengineered foods. In addition, manufacturers should consider appropriate recordkeeping to document the segregation procedures to ensure that the food's labeling is not false or misleading. In some situations, certifications or affidavits from farmers, processors, and others in the food production and distribution chain may be adequate to document that foods are obtained from the use of traditional methods. A statement that a food is "free" of bioengineered material may be difficult to substantiate without testing. Because appropriately validated testing methods are not currently available for many foods, it is likely that it would be easier to document handling practices and procedures to substantiate a claim about how the food was processed than to substantiate a "free" claim.

Source: U.S. Food and Drug Administration. Center for Food Safety and Applied Nutrition. *Guidance for Industry: Voluntary Labeling Indicating Whether Foods Have or Have Not Been Developed Using Bioengineering.* January 2001. Available online. URL: http://www.cfsan.fda.gov/~dms/biolabgu.html. Accessed on March 19, 2009.

Regulations with Respect to Genetically Modified Foods: European Union (2003)

In 1997, the European Parliament adopted a set of rules and regulations controlling the use of genetically modified foods in the European

Community (EC No. 258/97). In 2003, the parliament adopted a number of revisions to those rules, covering the labeling and sale of foods and feed in Europe (EC No. 1829/2003). The directive is divided into four chapters. Chapter I deals with definitions and objectives of the act. Chapter II deals with genetically modified foods. Chapter III deals with genetically modified feeds. Chapter IV deals with additional administrative and regulatory provisions relating to either GM foods or feed, or both. Some sections of those rules are excerpted here.

CHAPTER II
Section 1
Article 4
Requirements

1. Food referred to in Article 3(1) must not:

(a) have adverse effects on human health, animal health or the environment;

(b) mislead the consumer;

(c) differ from the food which it is intended to replace to such an extent that its normal consumption would be nutritionally disadvantageous for the consumer.

2. No person shall place on the market a GMO for food use or food referred to in Article 3(1) unless it is covered by an authorisation granted in accordance with this Section and the relevant conditions of the authorisation are satisfied.

3. No GMO for food use or food referred to in Article 3(1) shall be authorised unless the applicant for such authorisation has adequately and sufficiently demonstrated that it satisfies the requirements of paragraph 1 of this Article.

*... [The rest of this article describes how an authorization is obtained for a GM food. Remaining articles in Section 1 (Articles 5–11) discuss further details about authorizing the sale of GM foods.]

Section 2
Labelling
Article 12
Scope

1. This Section shall apply to foods which are to be delivered as such to the final consumer or mass caterers in the Community and which:

(a) contain or consist of GMOs; or

(b) are produced from or contain ingredients produced from GMOs.

2. This Section shall not apply to foods containing material which contains, consists of or is produced from GMOs in a proportion no higher than 0,9 per cent of the food ingredients considered individually or food consisting of a single ingredient, provided that this presence is adventitious or technically unavoidable.

* ...

Article 13
Requirements

1. Without prejudice to the other requirements of Community law concerning the labelling of foodstuffs, foods falling within the scope of this Section shall be subject to the following specific labelling requirements:

(a) where the food consists of more than one ingredient, the words 'genetically modified' or 'produced from genetically modified (name of the ingredient)' shall appear in the list of ingredients provided for in Article 6 of Directive 2000/13/EC in parentheses immediately following the ingredient concerned;

(b) where the ingredient is designated by the name of a category, the words 'contains genetically modified (name of organism)' or 'contains (name of ingredient) produced from genetically modified (name of organism)' shall appear in the list of ingredients;

(c) where there is no list of ingredients, the words 'genetically modified' or 'produced from genetically modified (name of organism)' shall appear clearly on the labelling;

(d) the indications referred to in (a) and (b) may appear in a footnote to the list of ingredients. In this case they shall be printed in a font of at least the same size as the list of ingredients. Where there is no list of ingredients, they shall appear clearly on the labelling;

(e) where the food is offered for sale to the final consumer as non-pre-packaged food, or as pre-packaged food in small containers of which the largest surface has an area of less than 10 cm^2, the information required under this paragraph must be permanently and visibly displayed either on the food display or immediately next to it, or on the packaging material, in a font sufficiently large for it to be easily identified and read.* ...

[*Article 14 deals with more details about the labeling of products containing GM foods.*]

Source: "Regulation (EC) No 1829/2003 of the European Parliament and of the Council of 22 September 2003 on genetically modified food and feed." *Official Journal of the European Union*. 268 (October 18, 2003):

1–23. ©European Communities, http://eur-lex.europa.eu/. Only European Union legislation printed in the paper edition of the Official Journal of the European Union is deemed authentic. Accessed on March 19, 2009.

Alaska State Law on Genetic Privacy (2004)

Until 2008, there were only limited federal regulations dealing with genetic testing. By the time the first law was passed by the U.S. Congress (see next document), a number of states had adopted legislation regulating genetic testing in one way or another. The Alaska law, excerpted here, is an example of one way states have dealt with the issue of genetic testing.

Chapter 18.13. GENETIC PRIVACY
Sec. 18.13.010. Genetic testing.
  (a) Except as provided in (b) of this section,

    (1) a person may not collect a DNA sample from a person, perform a DNA analysis on a sample, retain a DNA sample or the results of a DNA analysis, or disclose the results of a DNA analysis unless the person has first obtained the informed and written consent of the person, or the person's legal guardian or authorized representative, for the collection, analysis, retention, or disclosure;

    (2) a DNA sample and the results of a DNA analysis performed on the sample are the exclusive property of the person sampled or analyzed.

  (b) The prohibitions of (a) of this section do not apply to DNA samples collected and analyses conducted

    (1) under AS 44.41.035 or comparable provisions of another jurisdiction;

    (2) for a law enforcement purpose, including the identification of perpetrators and the investigation of crimes and the identification of missing or unidentified persons or deceased individuals;

    (3) for determining paternity;

    (4) to screen newborns as required by state or federal law;

    (5) for the purpose of emergency medical treatment.

  (c) A general authorization for the release of medical records or medical information may not be construed as the informed and written consent required by this section. The Department of Health and Social Services may by regulation adopt a uniform informed and written consent form to assist persons in meeting the requirements of this section. A person using that uniform informed and written consent is exempt from civil or criminal liability for actions taken under the consent form. A person may revoke or amend their informed and written consent at any time.

Sec. 18.13.020. Private right of action.

A person may bring a civil action against a person who collects a DNA sample from the person, performs a DNA analysis on a sample, retains a DNA sample or the results of a DNA analysis, or discloses the results of a DNA analysis in violation of this chapter. In addition to the actual damages suffered by the person, a person violating this chapter shall be liable to the person for damages in the amount of $ 5,000 or, if the violation resulted in profit or monetary gain to the violator, $ 100,000.

Sec. 18.13.030. Criminal penalty.

(a) A person commits the crime of unlawful DNA collection, analysis, retention, or disclosure if the person knowingly collects a DNA sample from a person, performs a DNA analysis on a sample, retains a DNA sample or the results of a DNA analysis, or discloses the results of a DNA analysis in violation of this chapter.

(b) In this section, "knowingly" has the meaning given in AS 11.81.900.

(c) Unlawful DNA collection, analysis, retention, or disclosure is a class A misdemeanor.

Source: The Alaska Statutes. Available online. URL: http://www.legis.state.ak.us/cgi-bin/folioisa.dll/stattx07/query=*/doc/%7B@8125%7D?. Accessed on April 15, 2009.

Genetic Information Nondiscrimination Act (2008)

On January 16, 2007, Representative Louise Slaughter (D-NY) introduced into the U.S. House of Representatives legislation to protect individuals from discrimination based on personal information obtained as the result of genetic testing. That legislation worked its way through the U.S. Congress and was signed by President George W. Bush on May 21, 2008, becoming Public Law 110–233. The new law amends a number of existing laws dealing with employment, taxation, retirement, and related issues and also establishes a number of new regulations about the use of genetic testing by employers. Some important sections of that law follow.

SEC. 101. AMENDMENTS TO EMPLOYEE RETIREMENT INCOME SECURITY ACT OF 1974.
(a) NO DISCRIMINATION IN GROUP PREMIUMS BASED ON GENETIC INFORMATION.—
* . . .

"(A) IN GENERAL.—For purposes of this section, a group health plan, and a health insurance issuer offering group health insurance coverage in connection with a group health plan, may not adjust premium or contribution amounts for the group covered under such plan on the basis of genetic information.
* . . .

(b) LIMITATIONS ON GENETIC TESTING; PROHIBITION ON COLLECTION OF GENETIC INFORMATION; APPLICATION TO ALL PLANS.
* . . .

TESTING.—A group health plan, and a health insurance issuer offering health insurance coverage in connection with a group health plan, shall not request or require an individual or a family member of such individual to undergo a genetic test.
* . . .

"(d) PROHIBITION ON COLLECTION OF GENETIC INFORMATION.—

"(1) IN GENERAL.—A group health plan, and a health insurance issuer offering health insurance coverage in connection with a group health plan, shall not request, require, or purchase genetic information for underwriting purposes (as defined in section 733).
* . . . [*Section 102 deals with amendments to the Public Health Service Act. Section 103 deals with amendments to the Internal Revenue Service code. Section 104 deals with amendments to the Social Service Act. Sections 105 and 106 are "housekeeping" sections dealing with privacy, confidentiality, and related matters.*]"

TITLE II—PROHIBITING EMPLOYMENT DISCRIMINATION ON THE BASIS OF GENETIC INFORMATION
[Section 201 supplies definitions of terms used in this title.]
SEC. 202. EMPLOYER PRACTICES.

(a) DISCRIMINATION BASED ON GENETIC INFORMATION.—It shall be an unlawful employment practice for an employer—

(1) to fail or refuse to hire, or to discharge, any employee, or otherwise to discriminate against any employee with respect to the compensation, terms, conditions, or privileges of employment of the employee, because of genetic information with respect to the employee; or

(2) to limit, segregate, or classify the employees of the employer in any way that would deprive or tend to deprive any employee of employment opportunities or otherwise adversely affect the status of the employee as an employee, because of genetic information with respect to the employee.

(b) ACQUISITION OF GENETIC INFORMATION.—It shall be an unlawful employment practice for an employer to request, require, or purchase genetic information with respect to an employee or a family member of the employee except—

(1) where an employer inadvertently requests or requires family medical history of the employee or family member of the employee;

(2) where—

(A) health or genetic services are offered by the employer, including such services offered as part of a wellness program;

(B) the employee provides prior, knowing, voluntary, and written authorization;

(C) only the employee (or family member if the family member is receiving genetic services) and the licensed health care professional or board certified genetic counselor involved in providing such services receive individually identifiable information concerning the results of such services; and

(D) any individually identifiable genetic information provided under subparagraph (C) in connection with the services provided under subparagraph (A) is only available for purposes of such services and shall not be disclosed to the employer except in aggregate terms that do not disclose the identity of specific employees;

(3) where an employer requests or requires family medical history from the employee to comply with the certification provisions of section 103 of the Family and Medical Leave Act of 1993 (29 U.S.C. 2613) or such requirements under State family and medical leave laws;

(4) where an employer purchases documents that are commercially and publicly available (including newspapers, magazines, periodicals, and books, but not including medical databases or court records) that include family medical history;

(5) where the information involved is to be used for genetic monitoring of the biological effects of toxic substances in the workplace, but only if—

(A) the employer provides written notice of the genetic monitoring to the employee;

(B)

(i) the employee provides prior, knowing, voluntary, and written authorization; or

(ii) the genetic monitoring is required by Federal or State law;

(C) the employee is informed of individual monitoring results;

(D) the monitoring is in compliance with—

(i) any Federal genetic monitoring regulations, including any such regulations that may be promulgated by the Secretary of Labor pursuant to the Occupational Safety and Health Act of 1970 (29 U.S.C.651 et seq.), the Federal Mine Safety and Health Act of 1977 (30 U.S.C. 801 et seq.), or the Atomic Energy Act of 1954 (42 U.S.C. 2011 et seq.); or

(ii) State genetic monitoring regulations, in the case of a State that is implementing genetic monitoring regulations under the authority of the Occupational Safety and Health Act of 1970 (29 U.S.C. 651 et seq.); and

(E) the employer, excluding any licensed health care professional or board certified genetic counselor that is involved in the genetic monitoring program, receives the results of the monitoring only in aggregate terms that do not disclose the identity of specific employees; or

(6) where the employer conducts DNA analysis for law enforcement purposes as a forensic laboratory or for purposes of human remains identification, and requests or requires genetic information of such employer's employees, but only to the extent that such genetic information is used for analysis of DNA identification markers for quality control to detect sample contamination.

* . . . [*Section 203 deals with requirements of employment agencies with regard to genetic testing.*]

SEC. 204. LABOR ORGANIZATION PRACTICES.

(a) DISCRIMINATION BASED ON GENETIC INFORMATION.—It shall be an unlawful employment practice for a labor organization—

(1) to exclude or to expel from the membership of the organization, or otherwise to discriminate against, any member because of genetic information with respect to the member;

(2) to limit, segregate, or classify the members of the organization, or fail or refuse to refer for employment any member, in any way that

would deprive or tend to deprive any member of employment opportunities, or otherwise adversely affect the status of the member as an employee, because of genetic information with respect to the member; or

(3) to cause or attempt to cause an employer to discriminate against a member in violation of this title.

* . . . [*The rest of this section provides restrictions on unions similar to those applied to employers in section 201. Section 205 deals with nondiscrimination in training programs. Section 206 deals with confidentiality of genetic information. Sections 207 through 210 and Title III deal with miscellaneous provisions and "housekeeping" issues related to enactment and enforcement of the law.*]

Source: Public Law 110–233—may 21, 2008. Available online. URL: http://frwebgate.access.gpo.gov/cgi-bin/getdoc.cgi?dbname=110 _cong_public_laws&docid=f:publ233.110.pdf. Accessed on March 17, 2009.

Guidance for APHIS Permits for Field Testing or Movement of Organisms Intended for Pharmaceutical or Industrial Use (2008)

One area of DNA technology that is highly regulated by the U.S. government is the testing or movement of genetically modified organisms designed for the production of pharmaceutical or industrial compounds. In July 2008, the Animal and Plant Health Inspection Service (APHIS) of the U.S. Department of Agriculture issued a detailed (42-page) set of instructions about the steps to be followed in such a practice. Table 6.2 summarizes the steps that must be taken both by the applicant and by the APHIS before a permit is issued for research in the field.

State Laws

In contrast to the dearth of federal legislation on DNA technologies (in general), a rather large number of laws have been passed by individual states on the technology. These laws tend to differ widely from each other, some, for example, outlawing both reproductive and therapeutic cloning, and others allowing therapeutic cloning, but not reproductive

TABLE 6.2

Overview of Applicant Activities and APHIS Oversight of the Environmental Release of Regulated Articles

Applicant	APHIS
• Optional early consultation with APHIS	• Optional early consultation with Applicant
• Familiarize with pertinent regulations and APHIS guidance documents	• Discussions with applicant on procedures and processes required for proposed field test
• Gather information requested in permit application	• Receive permit application and review for administrative completeness
• Determine environmental release site location	• Request further information for administrative completeness
• Design confinement conditions for environmental release	• Assign Biotechnologist for scientific review
• Write Standard Operating Procedures (SOPs)	• Scientific Review of permit, SOPs and training of personnel
• Identify employees and training required for those employees	• Request changes to the permit application or procedures as required by APHIS
• Submit Permit at least 120 days prior to release if no Environmental Assessment (EA) is required, 180 days if EA is required	• Request consultations if necessary
• Provide addition information requested by APHIS	• Determine if environmental release requires Environmental Assessment (EA) or Environmental Impact Statement (EIS), before permit can be issued
• Revise procedures, SOPs and training as requested by APHIS	• If EA is required, write EA, publish EA in the Federal Register for Public Comment, respond to public comments and determine if APHIS can reach a Finding of No Significant Impact (FONSI)
• Train Employees	• Send letter to state notifying of permit application
• Receive Approved Permit with attached APHIS permit and APHIS supplemental permit conditions	• Receive state response and work with state if issues are raised
• Complete training of employees and train any new employees during the environmental release	• Determine if permit can be issued and if so issues permit
• Keep records of training	• Inspect field trial site at all critical control points before, during, and after release of the regulated article
• Submit reports to APHIS during the environmental release as specified in the APHIS permit conditions and APHIS supplemental permit conditions	• Review records and observe field trial operations to assure that all permit conditions are met including, training, planting, monitoring, cleaning, reproductive control, harvest, and movement
• Conduct environmental release as described in the regulations, permit, APHIS permit conditions, APHIS supplemental permit conditions and SOPs	• Follow up on all noncompliance incidents and review corrective actions implemented to assure compliance with all Federal Regulations
• Keep records of actions during the environmental release as specified in the permit, permit conditions and supplemental permit conditions	• Review reports or records of potential harmful effects on the environment
• Monitor and keep records of monitoring the environmental release site for volunteers any new employees during the environmental release	• Determine that permit conditions for devitalization and disposal of the regulated article are fully met
	• Review reports submitted by permitee

Source: Executive Office of the President. Office of Science and Technology Policy. "Coordinated Framework for Regulation of Biotechnology." *Federal Register,* June 26, 1986, 23302. Available online. URL: http:// usbiotechreg.nbii.gov/CoordinatedFrameworkForRegulationOfBiotechnology1986.pdf.

cloning. For a good review of laws and legislation on DNA technology, see National Conference of State Legislatures, "State Genetics Laws," online at http://www.ncsl.org/Default.aspx?TabID=160&tabs= 832,126,900#832. An example of just one state law, that of Rhode Island, on a single topic, cloning, is provided here.

TITLE 23
Health and Safety
CHAPTER 23-16.4
Human Cloning
SECTION 23-16.4-1

§ 23-16.4-1 Declaration of intent and purpose.—Whereas, recent medical and technological advances have had tremendous benefit to patients, and society as a whole, and biomedical research for the purpose of scientific investigation of disease or cure of a disease or illness should be preserved and protected and not be impeded by regulations involving the cloning of an entire human being; and

Whereas, molecular biology, involving human cells, genes, tissues, and organs, has been used to meet medical needs globally for twenty (20) years, and has proved a powerful tool in the search for cures, leading to effective medicines to treat cystic fibrosis, diabetes, heart attack, stroke, hemophilia, and HIV/AIDS;

The purpose of this legislation is to place a ban on the creation of a human being through division of a blastocyst, zygote, or embryo or somatic cell nuclear transfer, and to protect the citizens of the state from potential abuse deriving from cloning technologies. This ban is not intended to apply to the cloning of human cells, genes, tissues, or organs that would not result in the replication of an entire human being. Nor is this ban intended to apply to in vitro fertilization, the administration of fertility enhancing drugs, or other medical procedures used to assist a woman in becoming or remaining pregnant, so long as that procedure is not specifically intended to result in the gestation or birth of a child who is genetically identical to another conceptus, embryo, fetus, or human being, living or dead.

* . . . [*This section is followed by three additional sections dealing with a prohibition on human cloning, penalties, and a sunset clause for the act.*]

Source: State of Rhode Island General Laws. Chapter 23-16.4. Human Cloning. Index Of Sections. Available online. URL: http://www .rilin.state.ri.us/Statutes/TITLE23/23-16.4/INDEX.HTM. Accessed on March 17, 2009.

Declarations and Position Statements

United Nations Declaration on Human Cloning (2005)

The announcement of the first cloned mammal (the sheep, Dolly) in 1997 fueled a firestorm of concern about the possible use of DNA technology to clone a human being. One response evoked by this event was the request by some members of the United Nations General Assembly for a study of this issue and a statement as to the UN's position on human reproductive cloning. The first formal request for such an action was presented in 2001, and a committee to study the issue and publish its report was appointed by the General Assembly in that year. Committee negotiations did not go smoothly for a number of reasons, most important of which was that some nations wished to ban all forms of cloning, including both therapeutic and reproductive cloning, while other nations preferred a limited ban on reproductive cloning only. This division was reflected in the final vote on the declaration adopted by the General Assembly on March 8, 2005, by a vote of 84 to 34, with 37 abstentions and 36 members being absent from the vote. The text of that declaration follows.

United Nations Declaration on Human Cloning

The General Assembly,

    Guided by the purposes and principles of the Charter of the United Nations,

    Recalling the Universal Declaration on the Human Genome and Human Rights, adopted by the General Conference of the United Nations Educational, Scientific and Cultural Organization on 11 November 1997, and in particular article 11 thereof, which states that practices which are contrary to human dignity, such as the reproductive cloning of human beings, shall not be permitted,

    Recalling also its resolution 53/152 of 9 December 1998, by which it endorsed the Universal Declaration on the Human Genome and Human Rights,

    Aware of the ethical concerns that certain applications of rapidly developing life sciences may raise with regard to human dignity, human rights and the fundamental freedoms of individuals,

    Reaffirming that the application of life sciences should seek to offer relief from suffering and improve the health of individuals and humankind as a whole,

Emphasizing that the promotion of scientific and technical progress in life sciences should be sought in a manner that safeguards respect for human rights and the benefit of all,

Mindful of the serious medical, physical, psychological and social dangers that human cloning may imply for the individuals involved, and also conscious of the need to prevent the exploitation of women,

Convinced of the urgency of preventing the potential dangers of human cloning to human dignity,

Solemnly declares the following:

(a) Member States are called upon to adopt all measures necessary to protect adequately human life in the application of life sciences;

(b) Member States are called upon to prohibit all forms of human cloning inasmuch as they are incompatible with human dignity and the protection of human life;

(c) Member States are further called upon to adopt the measures necessary to prohibit the application of genetic engineering techniques that may be contrary to human dignity;

(d) Member States are called upon to take measures to prevent the exploitation of women in the application of life sciences;

(e) Member States are also called upon to adopt and implement without delay national legislation to bring into effect paragraphs (a) to (d);

(f) Member States are further called upon, in their financing of medical research, including of life sciences, to take into account the pressing global issues such as HIV/AIDS, tuberculosis and malaria, which affect in particular the developing countries.

Source: United Nations General Assembly. "International Convention Against the Reproductive Cloning of Human Beings." A/59/516/ Add. 1, February 24, 2005. © United Nations, 2005. Reproduced with permission.

Additional Protocol to the Convention on Human Rights and Biomedicine, Concerning Genetic Testing for Health Purposes (2008)

This document has a long life, dating back to the late 1990s when the Steering Committee on Bioethics of the Council of Europe began deliberations on the use of genetic testing in the diagnosis of human disease. Those deliberations became very complex, and the committee's final report was not issued until almost a decade later. The final protocol consists of 11 chapters and 28 articles, whose major foci are as follows:

Chapter I—Object and scope
Article 1—Object and purpose
Parties to this Protocol shall protect the dignity and identity of all human beings and guarantee everyone, without discrimination, respect for their integrity and other rights and fundamental freedoms with regard to the tests to which this Protocol applies in accordance with Article 2.

Article 2—Scope
1 This Protocol applies to tests, which are carried out for health purposes, involving analysis of biological samples of human origin and aiming specifically to identify the genetic characteristics of a person which are inherited or acquired during early prenatal development (hereinafter referred to as "genetic tests").

2 This Protocol does not apply:

a to genetic tests carried out on the human embryo or foetus;

b to genetic tests carried out for research purposes. * . . .

Chapter II—General provisions
Article 3—Primacy of the human being
The interests and welfare of the human being concerned by genetic tests covered by this Protocol shall prevail over the sole interest of society or science.

Article 4—Non-discrimination and non-stigmatisation
1 Any form of discrimination against a person, either as an individual or as a member of a group on grounds of his or her genetic heritage is prohibited.

2 Appropriate measures shall be taken in order to prevent stigmatisation of persons or groups in relation to genetic characteristics.

Chapter III—Genetic services
Article 5—Quality of genetic services
Parties shall take the necessary measures to ensure that genetic services are of appropriate quality. In particular, they shall see to it that:

a genetic tests meet generally accepted criteria of scientific validity and clinical validity;

b a quality assurance programme is implemented in each laboratory and that laboratories are subject to regular monitoring;

c persons providing genetic services have appropriate qualifications to enable them to perform their role in accordance with professional obligations and standards.

Article 6—Clinical utility
Clinical utility of a genetic test shall be an essential criterion for deciding to offer this test to a person or a group of persons.

Article 7—Individualised supervision
1 A genetic test for health purposes may only be performed under individualised medical supervision.
2 Exceptions to the general rule referred to in paragraph 1 may be allowed by a Party, subject to appropriate measures being provided, taking into account the way the test will be carried out, to give effect to the other provisions of this Protocol.

However, such an exception may not be made with regard to genetic tests with important implications for the health of the persons concerned or members of their family or with important implications concerning procreation choices.

Chapter IV—Information, genetic counselling and consent
Article 8—Information and genetic counselling
1 When a genetic test is envisaged, the person concerned shall be provided with prior appropriate information in particular on the purpose and the nature of the test, as well as the implications of its results.
2 For predictive genetic tests as referred to in Article 12 of the Convention on Human Rights and Biomedicine, appropriate genetic counselling shall also be available for the person concerned.
The tests concerned are:

- tests predictive of a monogenic disease,
- tests serving to detect a genetic predisposition or genetic susceptibility to a disease,
- tests serving to identify the subject as a healthy carrier of a gene responsible for a disease.

The form and extent of this genetic counselling shall be defined according to the implications of the results of the test and their significance for the person or the members of his or her family, including possible implications concerning procreation choices. Genetic counselling shall be given in a non-directive manner.

Article 9—Consent
1 A genetic test may only be carried out after the person concerned has given free and informed consent to it.
Consent to tests referred to in Article 8, paragraph 2, shall be documented.
2 The person concerned may freely withdraw consent at any time.

* . . . [*Chapter V deals with genetic testing of individuals unable to give their consent. Chapter VI deals with genetic testing of family members who are unable to give consent, who can not be contacted, or who are deceased.*]

Chapter VII—Private life and right to information
Article 16—Respect for private life and right to information
1 Everyone has the right to respect for his or her private life, in particular to protection of his or her personal data derived from a genetic test.
2 Everyone undergoing a genetic test is entitled to know any information collected about his or her health derived from this test. The conclusions drawn from the test shall be accessible to the person concerned in a comprehensible form.
3 The wish of a person not to be informed shall be respected.
4 In exceptional cases, restrictions may be placed by law on the exercise of the rights contained in paragraphs 2 and 3 above in the interests of the person concerned.
Chapter VIII—Genetic screening programmes for health purposes
Article 19—Genetic screening programmes for health purposes
A health screening programme involving the use of genetic tests may only be implemented if it has been approved by the competent body. This approval may only be given after independent evaluation of its ethical acceptability and fulfilment of the following specific conditions:
  a the programme is recognised for its health relevance for the whole population or section of population concerned;
  b the scientific validity and effectiveness of the programme have been established;
  c appropriate preventive or treatment measures in respect of the disease or disorder which is the subject of the screening, are available to the persons concerned;
  d appropriate measures are provided to ensure equitable access to the programme;
  e the programme provides measures to adequately inform the population or section of population concerned of the existence, purposes and means of accessing the screening programme as well as the voluntary nature of participation in it.

* . . . [*Chapter IX deals with public information programs about genetic testing. Chapter X deals with the relation between this protocol and the general Convention. Chapter XI deals with "housekeeping" issues needed to carry out terms of the protocol.*]

Source: Additional Protocol to the Convention on Human Rights and Biomedicine, concerning Genetic Testing for Health Purposes. Strasbourg: Council of Europe, November 27, 2008. Available online. URL: http://conventions.coe.int/Treaty/EN/Treaties/Html/203.htm. Accessed on March 18, 2009.

Human Cloning: The Need for A Comprehensive Ban (2001)

The Center for Bioethics and Human Dignity (CBHD) was formed in 1993 by a group of about a dozen Christian ethicists who felt that the Christian perspective on bioethical issues had not been receiving adequate attention. Since that time, CBHD scholars have prepared position papers on a number of issues, including human cloning. The center's paper on this topic concludes with the section that follows.

VI. Conclusion.
Because the prospect of human cloning carries great potential to impact humanity in ways previously only imagined, it is exceedingly important that Congress adopt legislation that will protect society and the citizens who live in it—both now and for generations to come. We believe that the following points are of primary significance to the current legislative debate on this issue:

I.   The overwhelming consensus in this country that human reproductive cloning should not be permitted necessitates a ban on both reproductive and "therapeutic" cloning.

II.   To mandate the destruction of clonal human embryos created for research purposes would constitute a break with our nation's longstanding legal tradition and much of public sentiment.

III. The United States should promote ethical scientific and medical research, and not merely the progress of research, as "good ends" do not justify any and all means to achieve those ends.

IV.   The pursuit of therapies for human disease and disability via "therapeutic" cloning would likely leave many Americans without acceptable means to relieve their suffering.

V.   Human beings have a right not to be created for purposes of experimentation.

It is our contention that careful consideration of these points leads to support for a comprehensive ban prohibiting both the reproductive and "therapeutic" cloning of human beings. Failure to adopt such a ban will result in scientific, ethical, and legal failures—the scope and consequences of which will be of great magnitude.

Source: "Position Statement: Human Cloning: The Need for a Comprehensive Ban." URL: http://www.cbhd.org/content/cloning-position-statement. Accessed on August 22, 2009. The contents of this article do not necessarily reflect the opinions of CBHD, its staff, board or supporters. Permission to reprint granted as long as The Center for Bioethics and Human Dignity and the Web address for this article is referenced.

BIO Statement of Ethical Principles (2008)

BIO is an acronym for the Biotechnology Industry Organization, the world's largest biotechnology organization. It provides advocacy, business development, and communications services for its more than 1,200 members. In 1997, the organization's board of directors adopted a statement of ethical principles, which it reprints in its annual report. The most recent expression of that statement, given below, is from its Guide to Biotechnology 2008. *The following excerpt lists the principles only. The complete document also includes further explication of each of these principles.*

We respect the power of biotechnology and apply it for the benefit of humankind.

We listen carefully to those who are concerned about the implications of biotechnology and respond to their concerns.

We help educate the public about biotechnology, its benefits and implications.

We place our highest priority on health, safety and environmental protection in the use of our products.

We support strong protection of the confidentiality of medical information, including genetic information.

We respect the animals involved in our research and treat them humanely.

We are sensitive to and considerate of the ethical and social issues regarding genetic research.

We adhere to strict informed-consent procedures.

We develop our agricultural products to enhance the world's food supply and to promote sustainable agriculture with attendant environmental benefits.

We develop environmental biotechnology to clean up hazardous waste more efficiently with less disruption to the environment and to prevent pollution by treating waste before it is released.

We oppose the use of biotechnology to develop weapons.

We continue to support the conservation of biological diversity.

We will abide by the ethical standards of the American Medical Association and, where appropriate, other health-care professional societies to ensure that our products are appropriately prescribed, dispensed and used.

Source: Biotechnology Industry Organization. *Guide to Biotechnology 2008*. Washington, D.C.: Biotechnology Industry Organization. Used by permission.

Reports

Report and Recommendations of the Panel to Assess the NIH Investment in Research on Gene Therapy (1995)

By the mid-1990s, the potential applications of recombinant DNA research to a variety of fields were becoming more clearly apparent to policy makers around the world. In the United States, this awareness led to a number of policy studies about the possible future of rDNA research and its social and ethical implications. One of the most important of those studies was one commissioned by Harold Varmus, then director of the U.S. National Institutes of Health. The recommendations made by the authors of this report, although modified on a number of occasions, have expressed some fundamental views about the use of gene therapy in the treatment of human diseases.

The Panel finds that:
1. Somatic gene therapy is a logical and natural progression in the application of fundamental biomedical science to medicine and offers extraordinary potential, in the long-term, for the management and correction of human disease, including inherited and acquired disorders, cancer, and AIDS. The concept that gene transfer might be used to treat disease is founded on the remarkable advances of the past two decades in recombinant DNA technology. The types of diseases under consideration for gene therapy are diverse; hence, many different treatment strategies are being investigated, each with its own set of scientific and clinical challenges.
2. While the expectations and the promise of gene therapy are great, clinical efficacy has not been definitively demonstrated at this time in any gene therapy protocol, despite anecdotal claims of successful therapy and the initiation of more than 100 Recombinant DNA Advisory Committee (RAC)-approved protocols.
3. Significant problems remain in all basic aspects of gene therapy. Major difficulties at the basic level include shortcomings in all current

gene transfer vectors and an inadequate understanding of the biological interaction of these vectors with the host.

4. In the enthusiasm to proceed to clinical trials, basic studies of disease pathophysiology, which are likely to be critical to the eventual success of gene therapy, have not been given adequate attention. Such studies can lead to better definition of the important target cell(s) and to more effective design of the therapeutic approach. They often can be carried out in appropriate animal models. Pathophysiologic studies may also suggest alternative treatment strategies.

5. There is a clear and legitimate need for clinical studies to evaluate various aspects of gene therapy approaches. Although animal investigations are often valuable, it is not always possible to extrapolate directly from animal experiments to human studies. Indeed, in some cases, such as cystic fibrosis, cancer, and AIDS, animal models do not satisfactorily mimic the major manifestations of the corresponding human disease. Clinical studies represent not only practical implementation of basic discoveries, but also critical experiments which refine and define new questions to be addressed by non-clinical investigation.

6. Interpretation of the results of many gene therapy protocols has been hindered by a very low frequency of gene transfer, reliance on qualitative rather than quantitative assessments of gene transfer and expression, lack of suitable controls, and lack of rigorously defined biochemical or disease endpoints. The impression of the Panel is that only a minority of clinical studies, illustrated by some gene marking experiments, have been designed to yield useful basic information.

7. Overselling of the results of laboratory and clinical studies by investigators and their sponsors—be they academic, federal, or industrial—has led to the mistaken and widespread perception that gene therapy is further developed and more successful than it actually is. Such inaccurate portrayals threaten confidence in the integrity of the field and may ultimately hinder progress toward successful application of gene therapy to human disease.

Based on these findings, the Panel recommends the following:

1. In order to confront the major outstanding obstacles to successful somatic gene therapy, greater focus on basic aspects of gene transfer, and gene expression within the context of gene transfer approaches, is required. Such efforts need to be applied to improving vectors for gene delivery, enhancing and maintaining high level expression of genes transferred to somatic cells, achieving tissue-specific and regulated expression of transferred genes, and directing gene transfer to specific cell types. To stimulate innovative research, the Panel recommends the use of interdisciplinary workshops, specific program announcements in these areas, and the use of short-term, pilot grants for testing new

ideas and for encouraging investigators from other areas to enter the field of gene therapy.

2. To address important biological questions and provide a basis for the discovery of alternative treatment modalities, the Panel recommends increased emphasis on research dealing with the mechanisms of disease pathogenesis, further development of animal models of disease, enhanced use of preclinical gene therapy approaches in these models, and greater study of stem cell biology in diverse organ systems.

3. Strict adherence to high standards for excellence in clinical protocols must be demanded of investigators. Gene therapy protocols need to meet the same high standards required for all forms of translational (or clinical) research, whatever the enthusiasm for this (or any other) treatment approach.

4. To enhance the overall level of research in this area, the Panel recommends that NIH support broad interdisciplinary postdoctoral training of M.D. and Ph.D. investigators at the interface of clinical and basic science. Mechanisms for physician training in this area might include use of career development awards based on a program announcement in gene therapy.

5. Investigators in the field and their supporters need to be more restrained in their public discussion of findings, publications, and immediate prospects for the successful implementation of gene therapy approaches. The Panel recommends a concerted effort on the part of scientists, clinicians, science writers, research advocates, research institutions, industry, and the press to inform the public about not only the extraordinary promise of gene therapy, but also its current limitations.

6. NIH has already provided an appropriate initial investment in gene therapy. Future gene therapy research should compete with other forms of biomedical research for funding under stringent peer review. Only with fair, yet critical, peer review will high standards be met and maintained. The Panel specifically does not recommend special gene therapy study sections, expansion of existing center programs in gene therapy, or expansion of the recently funded core vector production program. To ensure that the level of support remains appropriate, the NIH investment in this field should be reexamined periodically.

7. To enhance the contribution of industry to the field, the Panel recommends that NIH encourage collaborative arrangements between academic institutions and industry that complement NIH-supported research, and also implement mechanisms that facilitate the distribution and testing of vectors and adjunct materials for use in clinical studies.

8. In an effort to improve gene therapy research and reduce duplication of effort, the Panel urges better coordination and scientific

review of such research throughout the NIH Intramural Program. In addition, NIH Institute Directors should resist pressures to include gene therapy research in their portfolios (either Intramural or Extramural) to "round out" their programs or compete with other Institutes. Instead, they should include such research only when there are compelling scientific reasons to go forward. Institute Directors should take the lead, where it seems appropriate, to focus efforts on improvement of diagnosis and understanding of disease pathogenesis and await further developments in vector technology before expanding clinical gene therapy programs.

Source: Orkin, Stuart H., and Arno G. Motulsky. *Report and Recommendations of the Panel to Assess the Nih Investment in Research on Gene Therapy*, December 7, 1995. Available online. URL: http:// www.nih.gov/news/panelrep.html. Accessed on March 19, 2009.

Promoting Safe and Effective Genetic Testing in the United States (1997)

The Human Genome Project was a massive research effort that ran from 1990 to 2003 with the goal of identifying and characterizing the 20,000–25,000 genes that make up the human genome. Along with this research effort, the program had a number of other objectives, including a study of the ethical, legal, and social issues (ELSI) that might arise as a result of the project. In 1994, a committee especially concerned with this issue, the National Institutes of Health–Department of Energy Working Group on Ethical, Legal and Social Implications of Human Genome Research identified a number of issues related to genetic testing that it felt required special attention and appointed a committee to review those issues and make recommendations about possible government action on them. That committee, the Task Force on Genetic Testing, issued its report in September 1997. The recommendations made by the task force are numerous and distributed throughout its report. The following excerpt includes the "overarching principles" the task force followed in developing its recommendations and a few of its most important recommendations.

Overarching Principles

In making recommendations on safety and effectiveness, the Task Force concentrated on test validity and utility, laboratory quality, and provider competence. It recognizes, however, that other issues impinge on testing, and problems may arise from testing. Regarding these issues, the Task Force endorses the following principles.

Informed Consent. The Task Force strongly advocates written informed consent. The failure of the Task Force to comment on informed consent for other uses does not imply that it should not be obtained.

Test Development. Informed consent for any validation study must be obtained whenever the specimen can be linked to the subject from which it came.

Testing in Clinical Practice. (1) It is unacceptable to coerce or intimidate individuals or families regarding their decision about predictive genetic testing. Respect for personal autonomy is paramount. People being offered testing must understand that testing is voluntary. Their informed consent should be obtained. Whatever decision they make, their care should not be jeopardized.

(2) Prior to the initiation of predictive testing in clinical practice, health care providers must describe the features of the genetic test, including potential consequences, to potential test recipients.

Newborn Screening. (1) If informed consent is waived for a newborn screening test, the analytical and clinical validity and clinical utility of the test must be established, and parents must be provided with suffi-cient information to understand the reasons for screening. By clinical utility, the Task Force means that interventions to improve the outcome of the infant identified by screening have been proven to be safe and effective.

(2) For those disorders for which newborn screening is available but the tests have not been validated or shown to have clinical utility, written parental consent is required prior to testing.

Prenatal and Carrier Testing. Respect for an individual's/couples' beliefs and values concerning tests undertaken for assisting reproductive deci-sions is of paramount importance and can best be maintained by a non-directive stance. One way of ensuring that a non-directive stance is taken and that parents' decisions are autonomous, is through requiring informed consent.

Testing of Children. Genetic testing of children for adult onset diseases should not be undertaken unless direct medical benefit will accrue to the child and this benefit would be lost by waiting until the child has reached adulthood.

Confidentiality. Protecting the confidentiality of information is essential for all uses of genetic tests. (1) Results should be released only to those individuals for whom the test recipient has given consent for informa-tion release. Means of transmitting information should be chosen to minimize the likelihood that results will become available to unauthor-ized persons or organizations. Under no circumstances should results with identifiers be provided to any outside parties, including employers, insurers, or government agencies, without the test recipient's written consent. (2) Health care providers have an obligation to the person being tested not to inform other family members without the permission of the person tested, except in extreme circumstances.

Discrimination. No individual should be subjected to unfair discrimination by a third party on the basis of having had a genetic test or receiving an abnormal genetic test result. Third parties include insurers, employers, and educational and other institutions that routinely inquire about the health of applicants for services or positions.

Consumer Involvement in Policy Making. Although other stakeholders are concerned about protecting consumers, they cannot always provide the perspective brought by consumers themselves, the end users of genetic testing. Consumers should be involved in policy (but not necessarily in technical) decisions regarding the adoption, introduction, and use of new, predictive genetic tests.

Recommendations

Need for an Advisory Committee on Genetic Testing

The Task Force calls on the Secretary of Health and Human Services (HHS) to establish an advisory committee on genetic testing in the Office of the Secretary. Members of the committee should represent the stakeholders in genetic testing, including professional societies (general medicine, genetics, pathology, genetic counseling), the biotechnology industry, consumers, and insurers, as well as other interested parties. The various HHS agencies with activities related to the development and delivery of genetic tests should send nonvoting representatives to the advisory committee, which can also coordinate the relevant activities of these agencies and private organizations. The Task Force leaves it to the Secretary to determine the relationship of this advisory committee to others that may be created in the broader area of genetics and public policy, of which genetic testing is only one part.

The committee would advise the Secretary on implementation of recommendations made by the Task Force in this report to ensure that (a) the introduction of new genetic tests into clinical use is based on evidence of their analytical and clinical validity, and utility to those tested; (b) all stages of the genetic testing process in clinical laboratories meet quality standards; (c) health providers who offer and order genetic tests have sufficient competence in genetics and genetic testing to protect the well-being of their patients; and (d) there be continued and expanded availability of tests for rare genetic diseases.

The Task Force recognizes the widely inclusive nature of genetic tests. It is therefore essential that the advisory committee recommend policies for the Secretary's consideration by which agencies and organizations implementing recommendations can determine those genetic tests that need stringent scrutiny. Stringent scrutiny is indicated when a test has the ability to predict future inherited disease in healthy or

apparently healthy people, is likely to be used for that purpose, and when no confirmatory test is available. The advisory committee or its designate should define additional indications.
*...

Ensuring the Safety and Effectiveness of New Genetic Tests

1. The genotypes to be detected by a genetic test must be shown by scientifically valid methods to be associated with the occurrence of a disease. The observations must be independently replicated and subject to peer review.
2. Analytical sensitivity and specificity of a genetic test must be determined before it is made available in clinical practice.
3. Data to establish the clinical validity of genetic tests (clinical sensitivity, specificity, and predictive value) must be collected under investigative protocols. In clinical validation, the study sample must be drawn from a group of subjects representative of the population for whom the test is intended. Formal validation for each intended use of a genetic test is needed.
4. Before a genetic test can be generally accepted in clinical practice, data must be collected to demonstrate the benefits and risks that accrue from both positive and negative results*...

Ensuring the Quality of Laboratories Performing Genetic Tests

No clinical laboratory should offer a genetic test whose clinical validity has not been established, unless it is collecting data on clinical validity under either an IRB [institutional review board]-approved protocol or conditional premarket approval agreement with FDA (one of the options presented in chapter 2. The service laboratory should justify and document the basis of decisions to put new tests into service. Regardless of where the test to be adopted was developed, clinical laboratory directors are responsible for ensuring the analytic validity of each genetic test their laboratory intends to offer before they make the test available for use in clinical practice (outside of an investigative protocol).

Before routinely offering genetic tests that have been clinically validated, a laboratory must conduct a pilot phase in which it verifies that all steps in the testing process are operating appropriately. If the pilot study reveals that the laboratory is not as competent as other laboratories in performing the test, or the test does not detect as many people with the genetic alteration as anticipated, the laboratory should not proceed to report patient-specific results without attempting to rectify the problems.
*...

Improving Providers' Understanding of Genetic Testing
Greater Public Knowledge of Genetics. A knowledge base on genetics and genetic testing should be developed for the general public*...

Undergraduate and Graduate Medical Education. The Task Force encourages the development of genetics curricula in medical school and residency training to enable all physicians to recognize inherited risk factors in patients and families and appreciate issues in genetic testing and the use of genetic services*...

Licensure and Certification. The likelihood that genetics will be covered in curricula will improve if relevant genetics questions are included in general licensure and specialty board certification examinations, and if correctly answering a proportion of the genetics questions is needed to attain a passing score.

Continuing Medical Education. The full beneficial effects of improving medical school and residency curricula in genetics will not be felt for many years. Consequently, improving the ability of providers currently in practice to offer and interpret genetic tests correctly is of paramount importance. In addition to the basic curricula already considered, the Task Force recommends that each specialty involved with the care of patients with disorders with genetic components should design its own curriculum for continuing education in genetics.

Administrators and other nonphysician personnel who triage patients and/or make coverage or reimbursement decisions, such as those in managed care organizations, should also have knowledge of the benefits and risks of genetic testing.

The Task Force endorses the recent establishment of a National Coalition for Health Professional Education in Genetics (NCHPEG) by the American Medical Association, the American Nurses Association, and the National Human Genome Research Institute. *...

Demonstrating Provider Competence. Hospitals and managed care organizations, on advice from the relevant medical specialty departments, should require evidence of competence before permitting providers to order predictive genetic tests defined as needing stringent scrutiny or to counsel about them. Periodic, systematic medical record review, with feedback to providers, should also be used to ensure appropriate use of genetic tests. *...
*...

Genetic Testing for Rare Inherited Disorders
*... the development and maintenance of tests for rare genetic diseases must continue to be encouraged. A comprehensive system to collect data on rare diseases must be established *...
Dissemination of Information About Rare Diseases
Physicians who encounter patients with symptoms and signs of rare genetic diseases should have access to accurate information that will

enable them to include such diseases in their differential diagnosis, to know where to turn for assistance in clinical and laboratory diagnosis, and to locate laboratories that test for rare diseases. * . . .

Ensuring Continuity and Quality of Tests for Rare Diseases

To maintain and expand its database, ORD [Office of Rare Diseases] should identify laboratories worldwide that perform tests for rare genetic diseases, the methodology employed, and whether the tests they provide are in the investigational stage, or are being used for clinical diagnosis and decision making. * . . .

Ensuring the Quality of Genetic Tests for Rare Diseases

In accordance with current law, the Task Force recommends that any laboratory performing any genetic test on which clinical diagnostic and/or management decisions are made should be certified under CLIA [Clinical Laboratory Improvement Amendments]. Research laboratories that are not currently providing genetic test results to providers or patients but that plan to do so in the future must register under CLIA. Once a laboratory registers, it does not have to wait for a survey before performing clinical tests.

Research laboratories that provide physicians with results of genetic tests, which may be used for clinical decision making, must validate their tests and be subject to the same internal and external review as other clinical laboratories. Nevertheless, the proposed genetics subcommittee of CLIAC should consider developing regulatory language under the proposed genetics specialty that is less stringent, but does not sacrifice quality for laboratories that only occasionally and in small volume perform tests whose results are made available to health care providers or patients.

Source: Holtzman, Neil A., and Michael S. Watson. *Promoting Safe and Effective Genetic Testing in the United States: Final Report of the Task Force on Genetic Testing.* Available online. URL: http://www.genome.gov /10002393. Accessed on March 18, 2009.

Genetically Modified Foods (2002)

This report was written in response to a request by John E. Baldacci (D-ME) and John F. Tierney (D-MA), of the U.S. House of Representatives, who asked the General Accounting Office to identify possible risks posed by genetically modified foods, to review current procedures for monitoring GM foods in the United States, to estimate future developments in the production of GM foods, and to recommend regulatory changes that might be appropriate in light of these developments. The following section summarizes the main features of the agency's findings.

GM foods pose the same types of inherent risks to human health as conventional foods: they can contain allergens, toxins, and compounds known as antinutrients, which inhibit the absorption of nutrients. Before marketing a GM food, company scientists evaluate these risks—even though they are not routinely evaluated in conventional foods—to determine if the foods pose any heightened risks. While some GM foods have contained allergens, toxins, and antinutrients, the levels have been comparable to those foods' conventional counterparts. In evaluating GM foods, scientists perform a regimen of tests. Biotechnology experts whom we contacted agree that this regimen of tests is adequate in assessing the safety of GM foods. While some consumer groups, as well as some scientists from the European Union, have questioned the ethical or cultural appropriateness of genetically modifying foods, experts whom we contacted from these organizations also believe the tests are adequate for assessing the safety of these foods.

While FDA reports that its evaluation process includes the necessary controls for ensuring it obtains the safety data needed to evaluate GM foods, some biotechnology experts state that aspects of its evaluation process could be enhanced. FDA's controls include (1) communicating clearly—through the agency's 1992 policy statement and subsequent guidance—what safety data are necessary for its evaluations of GM food safety; (2) having teams of FDA experts in diverse disciplines evaluate company submissions for GM foods and request additional safety data, if necessary; and (3) tailoring the level of evaluation to match the degree of each submission's novelty, thereby assuring that staff have time to obtain necessary safety data. Nonetheless, FDA's overall evaluation process could be enhanced, according to some experts, by randomly verifying the test data that companies provide and by increasing the transparency of the evaluation process—including communicating more clearly the scientific rationale for the agency's final decision on a GM food safety assessment.

In the future, scientists generally expect that genetic modifications will increasingly change the composition of GM foods to enhance their nutritional value. For example, one company has modified a type of rice to contain beta-carotene. In countries where rice is a dietary staple, this rice may reduce the incidence of blindness caused by vitamin-A deficiency. Current tests have been adequate for evaluating the few GM foods with relatively simple compositional changes that FDA has reviewed so far. New testing technologies are being developed to evaluate the increasingly complex compositional changes expected. Some scientists view these new technologies as a potentially useful supplement for existing tests, while others believe that the technologies will offer a more comprehensive way of assessing the safety of all changes in GM foods.

Monitoring the long-term health risks of GM foods is generally nei-
ther necessary nor feasible, according to scientists and regulatory officials we contacted. In their view, such monitoring is unnecessary
because there is no scientific evidence, or even a hypothesis, suggesting
that long-term harm (such as increased cancer rates) results from these
foods. Furthermore, there is consensus among these scientists and
regulatory officials that technical challenges make long-term monitoring infeasible. Experts cite, for example, the technical inability to track
the health effects of GM foods separately from those of their conventional counterparts. A recent report by food and health organizations
affiliated with the United Nations also expresses skepticism about the
feasibility of identifying long-term health effects from GM foods.

*. . . [*In the conclusion to its report, the GAO made the recommendations that
follow.*]

To enhance FDA's safety evaluations of GM foods, we recommend that
the Deputy Commissioner of Food and Drugs direct the agency's
Center for Food Safety and Applied Nutrition to

- obtain, on a random basis, raw test data from companies,
  during or after consultations, as a means of verifying the
  completeness and accuracy of the summary test data
  submitted by companies; and
- expand its memos to file recording its decisions about
  GM foods to provide greater detail about its evaluations
  of the foods, including the level of evaluation provided,
  the similarity of the foods to foods previously evaluated,
  and the adequacy of the tests performed by the submitting companies.

Source: U.S. General Accounting Office. *Genetically Modified Foods.
Experts View Regimen of Safety Tests as Adequate, but FDA's Evaluation
Process Could Be Enhanced.* Washington, D.C.: U.S. General Accounting
Office, May 2002. Available online. URL: http://www.gao.gov/
new.items/d02566.pdf. Accessed on March 19, 2009.

The Changing Moral Focus of Newborn Screening (2008)

*The origins of this report date back to December 2005, when the Presi-
dent's Committee on Bioethics began a consideration of the relationship
of children and bioethics. One outcome of that discussion was a review
and analysis of the current state of newborn screening of children for*

genetic disorders in the United States. In this report, the committee reviews current practices and public policy on the screening of children, changes in principles and practices of screening, the future of screening, and the issue of mandatory versus elective screening programs. The committee concludes its report with recommendations for an ethical framework within which the practices of newborn screening can be placed.

IV. An Ethical Framework for the Ongoing Expansion of Newborn Screening: A Recommendation by the President's Council on Bioethics

* . . .

* . . . an ethically sound approach to public policy in newborn screening would, in the Council's opinion, include the following elements. It would:

1. Reaffirm the essential validity and continuing relevance of the classical Wilson-Jungner screening criteria. *[See below]*

2. Insist that mandatory newborn screening be recommended to the states only for those disorders that clearly meet the classical criteria. Such a disorder must pose a serious threat to the health of the child, its natural history must be well understood, and timely and effective treatment must be available, so that the intervention as a whole is likely to provide a substantial benefit to the affected child.

3. Endorse the view that screening for other conditions that fail to meet the classical criteria may be offered by the states to parents on a voluntary basis under a research paradigm. Such screening programs should be presented forthrightly as pilot studies, whose benefits and risks to the infant are not certain, and for which IRB [institutional review board] approval should be obtained in each state. A condition included in a pilot screening study should be moved to the mandatory screening panel only if the evidence clearly establishes that it now meets the classical criteria.

4. Affirm that, when differential diagnosis of some targeted disorders entails detection of other poorly understood conditions that would not otherwise be suitable candidates for newborn screening, such results need not be transmitted to the child's physician and parents. It should be left to the states to formulate rules governing whether and when to disclose those results.

5. Encourage the states to reach a consensus on a uniform panel of conditions clearly meriting mandatory screening. In contrast, diversity among the states in regard to the pilot conditions for which they offer optional screening is to be welcomed, as it permits the states to learn from each other's different experiences.

6. Urge a thorough and continuing re-evaluation of the disorders now recommended for inclusion in the mandatory screening panel, to ascertain whether they genuinely meet the classical criteria that would

justify mandatory screening of all newborns, or whether they instead are suitable candidates for pilot screening studies. In support of such continuing reevaluation, states should be encouraged to collect and share data on the short- and long-term outcomes for children who test positive for a genetic disorder, both those on the mandatory screening panel and those targeted by pilot programs.

7. Reject any simple application of the "technological imperative," i.e., the view that screening for a disorder is justified by the mere fact that it is detectable via multiplex assay, even if the disorder is poorly understood and has no established treatment. There should be no presumption that multiplex screening platforms are to be used in "full profile mode."

[*The ten Wilson-Jungner criteria mentioned here were developed by James Wilson and Gunnar Jungner in 1968 for a World Health Organization monograph on the principles and practices of screening for disease. They are as follows (from the current report).*]

1. The condition sought should be an important health problem.
2. There should be an accepted treatment for patients with recognized disease.
3. Facilities for diagnosis and treatment should be available.
4. There should be a recognizable latent or early symptomatic stage.
5. There should be a suitable test or examination.
6. The test should be acceptable to the population.
7. The natural history of the condition, including development from latent to declared disease, should be adequately understood.
8. There should be an agreed policy on whom to treat as patients.
9. The cost of case-finding (including diagnosis and treatment of patients diagnosed) should be economically balanced in relation to possible expenditure on medical care as a whole.
10. Case-finding should be a continuing process and not a "once and for all" project.

Source: *The Changing Moral Focus of Newborn Screening: An Ethical Analysis by the President's Council on Bioethics.* Washington, D.C., December 2008. Available online. URL: http://www.bioethics.gov/reports/newborn_screening/index.html. Accessed on March 19, 2009.

Data

Planting of Genetically Modified Crops in the United States, 2000–2008

The popularity of genetically modified crops in the United States increased significantly in the first decade of the twenty first century. Table 6.3 shows the increase in amount of crop acreage for the three most popular crops.

Global Area of Biotech Crops in 2007: by Country (I)

Biotech crops have grown significantly in popularity over the past decade or more. Figure 6.1 shows the growth in biotech crops from 1996 to 2007. In 2007, more than 100 million hectares (250 million acres) of such crops were in cultivation around the

TABLE 6.3
Planting of Genetically Modified Crops in the United States, 2000–2008

Percent of All Crops Planted that Were Genetically Modified									
Crop	2000	2001	2002	2003	2004	2005	2006	2007	2008
Corn	25	26	34	40	47	52	61	73	80
Cotton	61	69	71	73	76	79	83	87	86
Soybeans	54	68	75	81	85	87	89	91	92

Source: "Adoption of Genetically Engineered Crops in the U.S." URL: http://www.ers.usda.gov/Data/BiotechCrops/. Accessed on March 19, 2009.

FIGURE 6.1
Global Area of Biotech Crops

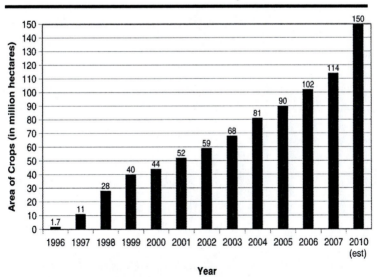

Source: International Service for the Acquisition of Agri-Biotech Applications. *Global Status of Commercialized Biotech/GM Crops: 2007.* URL: http://www.isaaa.org/resources/publications/briefs/37/executivesummary/default.html. Used by permission.

world. By some estimates, that number could reach 150 million hectares (375 million acres) by 2010.

Global Area of Biotech Crops in 2007: by Country (II)
Thirteen nations around the world are now sometimes classified as "biotech megacountries" because of the amount of farm acreage devoted to genetically engineered crops. Table 6.4 lists these

TABLE 6.4
Global Area of Biotech Crops in 2007: by Country (Million Hectares)

Rank	Country	Area (million hectares)	Biotech Crops
1	USA	57.7	Soybean, maize, cotton, canola, squash, papaya, alfalfa
2	Argentina	19.1	Soybean, maize, cotton
3	Brazil	6.2	Cotton
4	Canada	7.0	Canola, maize, soybean
5	India	6.2	Cotton
6	China	3.8	Cotton, tomato, poplar, petunia, papaya, sweet pepper
7	Paraguay	2.6	Soybean
8	South Africa	1.8	Maize, soybean, cotton
9	Uruguay	0.5	Soybean, maize
10	Philippines	0.3	Maize
11	Australia	0.1	Cotton
12	Spain	0.1	Maize
13	Mexico	0.1	Cotton, soybean
14	Colombia	<0.1	Cotton, carnation
15	Chile	<0.1	Maize, soybean, canola
16	France	<0.1	Maize
17	Honduras	<0.1	Maize
18	Czech Republic	<0.1	Maize
19	Portugal	<0.1	Maize
20	Germany	<0.1	Maize

Rank	Country	Area (million hectares)	Biotech Crops
21	Slovakia	<0.1	Maize
22	Romania	<0.1	Maize
23	Poland	<0.1	Maize

Source: International Service for the Acquisition of Agri-Biotech Applications. *Global Status of Commercialized Biotech/GM Crops: 2007* URL: http://www.isaaa.org/resources/publications/briefs/37/executivesummary/default.html. Used by permission.

13 countries, along with 10 more countries in which the growing of biotech crops has become significant.

Pharming Products

Although a fairly new technology, pharming has already resulted in the production of a relatively large number of useful products. Table 6.5 lists a sample of those products and the sources from which they come. This list of products and source plants is by no means complete.

TABLE 6.5
Useful Products Produced by Pharming Technologies

Product	Source Plant(s)
Angiotensin conversting enzyme (ACE)	Tobacco, tomato
Aprotinin (protein)	Maize
Canine gastric lipase (enzyme)	Tobacco
Cholera toxin B vaccine	Potato
Docosahexaenoic acid (omega-3 fatty acid)	Flax
Encephalin (protein)	Tobacco
Endostatin (protein fragment)	Tobacco
Epidermal growth factor	Tobacco
Gastric lipase (enzyme)	Corn
Hemoglobin (protein)	Tobacco
Hepatitis C vaccine	Potato, tobacco
Human α antitrypsin (protein)	Rice
Human acetylcholinesterase	Tomato
Human carcinoembryonic antigen	Tobacco
Human lactoferrin (protein)	Corn, rice, tobacco
Insulin (hormone)	Safflower

TABLE 6.5 continued

Product	Source Plant(s)
Interferon α, β (hormones)	Tobacco
Interferon γ (hormone)	Rice
Interleukin 2 (cytokin)	Potato
Interleukin 10, 12 (cytokins)	Tobacco
Norwalk virus vaccine	Potato, tobacco
Protein C (enzyme)	Tobacco
Rabbit virus protein	Tobacco
Somatotropin (human growth hormone)	Tobacco
Vascular endothelial growth factor	Moss

7

Organizations

A large number of organizations include some aspect of DNA technology in their work. Some of these organizations are responsible for carrying out legislative and regulatory guidelines for one or more types of genetic intervention; other organizations promote research on recombinant DNA issues; still other organizations campaign for or against some aspect of rDNA research and its applications. This chapter provides an introduction to some of the best known and most important of these organizations. They are arranged into two categories: governmental agencies, at both the international and U.S. federal levels, and nongovernmental organizations, which often are also advocacy groups working for or against some specific application of DNA technology.

Governmental Organizations

International

Ethics and Health
URL: http://www.who.int/ethics/en/

Ethics and Health is a section within the Department of Ethics, Equity, Trade, and Human Rights of the World Health Organization (WHO). Although the topic of the ethical issues related to the health profession and health treatments has long been a part of WHO's general program, the Ethics and Health section was created in October 2002 to deal specifically with many of these issues. Some topics on which the group has worked are ethical issues raised by the HIV/AIDS epidemic, long-term care, and human organ and tissue transplantation.

Publications: *Ethics, Access and Safety in Tissue and Organ Transplantation* (book/report); *Life Science Research: Opportunities and Risks for Public Health* (book/report); "A Dozen Questions (And Answers) on Human Cloning" (pamphlet/online page).

Food and Agriculture Organization (FAO)
URL: http://www.fao.org/

The Food and Agriculture Organization is an agency associated with the United Nations. Its mission is to help developing countries transition to modern agricultural, aquacultural, and fisheries practices and to promote good nutrition for all people of all nations. The organization's work is carried out through a number of departments, including Agriculture and Consumer Protection; Economic and Social Development; Fisheries and Aquaculture; Forestry; Knowledge and Communication; Natural Resources Management and Environment; Technical Cooperation; and Human, Financial and Physical Resources. The agency's work on issues related to biotechnology is generally not limited to a specific department, but is included as part of the work carried out within these general rubrics.

Publications: Many books, reports, newsletters, journals, press releases, and other print and electronic products. Also, *The State of The World's Animal Genetic Resources For Food And Agriculture* (book/report); "Biosafety Decisions and Perceived Commercial Risks" (discussion paper); *Genetically Modified Organisms in Crop Production and Their Effects on the Environment: Methodologies for Monitoring and The Way Ahead* (book/report); *Safety Assessment of Food Derived from Biotechnology* (report); "The Use of Reproductive and Molecular Biotechnology in Animal Genetic Resources Management—a Global Overview" (information bulletin); *Safety Assessment of Foods Derived from Recombinant-DNA Animals*.

International Bioethics Committee (IBC)
URL: http://portal.unesco.org/shs/en/ev.php-URL_ID=1879&URL_DO=DO_TOPIC&URL_SECTION=201.html

The International Bioethics Committee was created in 1993 to provide a forum for the study of developments in the life sciences and their potential implications for human dignity and freedom. The committee consists of 36 experts in the field who are appointed by the director general of the United Nations Educational, Scientific, and Cultural Organization (UNESCO) for a renewable term of four years. They come from a number of different fields, including medicine, genetics, chemistry, law, anthropology, philosophy, and history. The committee has drafted reports on a wide variety of topics ranging from intellectual property rights to plant biotechnology to genetic transformations in plants.

Publications: *Report of the IBC on Pre-implantation Genetic Diagnosis and Germ-line Intervention; Report on Confidentiality and Genetic Data; Food, Plant Biotechnology and Ethics; Report on Genetic Screening and Testing; Report on Human Gene Therapy; Genetic Counseling; Ethics and Neurosciences.*

International Centre for Genetic Engineering and Biotechnology (ICGEB)
URL: http://www.icgeb.org/

The International Centre for Genetic Engineering and Biotechnology was established in 1983 as the result of a treaty negotiated through the United Nations. Under the provisions of the treaty, the organization actually began operations in 1994 when the treaty was ratified by the required number of nations. At that point, the center became an autonomous intergovernmental agency. Its purpose is to conduct research in the field of genetic engineering and biotechnology that will have useful applications for developing nations, especially in the fields of biomedicine, crop improvement, environmental protection and remediation, biopharmaceuticals, and biopesticide production. The center originally had two components, one in Trieste, Italy, and one in New Delhi, India. In 2007, a third facility at Capetown, South Africa, was added. As of early 2009, 57 nations were members of the center, with an additional 20 nations awaiting ratification of the founding treaty. The center employs about 400 people from 38 different countries as research scientists, postdoctoral fellows, doctoral students, research technicians, and administrative personnel.

Publications: Over 1,700 research papers in a wide variety of professional journals.

International Criminal Police Organization (Interpol)
URL: http://www.interpol.int/

Interpol is an international law enforcement agency created in 1923 to facilitate cross-border cooperation between national criminal bureaus and to support all agencies worldwide whose mission it is to prevent or combat international crimes. The agency currently has 187 members. Its policies are set by a General Assembly and carried out by an Executive Secretary and a number of National Central Bureaus. The agency has a strong commitment to the use of DNA evidence in the investigation of criminal activity and has been charged by the General Assembly with providing "strategic and technical support to enhance member states' DNA profiling capacity and promote widespread use in the international law enforcement environment." The agency's DNA activities are operated by the Interpol DNA Unit Team, whose Web page is at http://www.interpol.int/public/forensic/dna/default.asp.

Publications: *Global DNA Inquiry: Results 2002* (report); "DNA Profiling" (fact sheet); *Interpol Handbook on DNA Data Exchange and Practice* (handbook).

United States

Armed Forces DNA Identification Laboratory (AFDIL)
URL: http://www.afip.org/Departments/oafme/dna/

The mission of the Armed Forces DNA Identification Laboratory is to provide research, consultation, and educational services to the U.S. Department of Defense and other federal agencies and to maintain and distribute, where needed, DNA specimens of U.S. military and other authorized personnel. The primary task of the agency is to identify members of the armed services who have been killed or lost in battle.

Publications: Forms and informational booklets dealing with the work of the agency.

Biotechnology Regulatory Services (BRS)
URL: http://www.aphis.usda.gov/biotechnology/brs_main.shtml

Biotechnology Regulatory Services is a division of the Animal Plant and Animal Health Inspection Service of the U.S. Department of Agriculture (USDA). The regulation of genetically modified organisms has been an issue of concern to the USDA since 1978 when the National Institutes of Health issued the first federal guidelines for the monitoring of genetically engineered organisms in the United States. Regulatory responsibility was assigned to a series of committees and working groups until 2002, when BRS was formed with exclusive responsibility for issuing permits, conducting inspections, and carrying out other activities required for the monitoring of genetically engineered organisms in agriculture. In 2008, the bureau published a comprehensive plan for the future of biotechnology in agriculture and its regulation in the United States.

Publications: A number of brochures, fact sheets, forms, newsletters, and other materials related to the bureau's work. Some topics included are "APHIS Biotechnology: Compliance with Regulations"; "APHIS Biotechnology: Contact Information for Alfalfa Farmers Regarding Roundup Ready Alfalfa"; "APHIS Biotechnology: Permitting Progress into Tomorrow"; "APHIS Publishes Request for Information on Genetically Engineered (GE) Animals, Q & A"; "Biotechnology, Federal Regulation, and the U.S. Department of Agriculture"; "Genetically Engineered Rice"; "National Environmental Policy Act and Its Role in USDA's Regulation of Biotechnology"; and "Permitting Genetically Engineered Plants That Produce Pharmaceutical Compounds."

Federal Bureau of Investigation (FBI)
URL: http://www.fbi.gov/

The Federal Bureau of Investigation is the primary federal agency responsible for investigating criminal activities in the United States. Its charge extends to all crimes not otherwise assigned to some other federal agency. The agency's CODIS Unit manages the Combined DNA Index System (CODIS) and the National DNA Index System (NDIS), the nation's largest collection of DNA specimens. The unit is also responsible for facilitating the interchange of DNA information and procedures among federal, state, and local crime laboratories and certain international law enforcement crime laboratories.

Publications: *CODIS: Combined DNA Index System* (brochure); *Quality Assurance Standards DNA Testing Laboratories* (brochure); *Quality Assurance Standards for DNA Databasing* (brochure).

Food and Drug Administration (FDA)
URL: http://www.fda.gov/

The Food and Drug Administration was created in 1862 as the Division of Chemistry in the U.S. Department of Agriculture. The agency took on its modern form after the passage in 1906 of the Federal Food and Drugs Act which, for the first time, assigned aggressive regulatory powers to the federal government in maintaining the purity of foods, drugs, and cosmetics in the United States. Today, the FDA has authority over almost all food products sold and consumed in the United States (other than meat and poultry); human and animal drugs; therapeutic agents of biological origin; medical devices; radiation-emitting products for consumer, medical, and occupational use; cosmetics; and animal feed. The agency's involvement with DNA technology and its social and ethical consequences is not limited to a single agency within the department, but is part of the mission of many divisions. For example, under its animal and veterinary responsibilities, the FDA has sponsored and conducted research and written reports and educational materials on issues such as animal cloning and genetically engineered animals.

Publications: *Regulation of Genetically Engineered Animals Containing Heritable Recombinant DNA Constructs* (guidance paper); "Genetically Engineered Animals" (fact sheet); "FDA Issues Final Guidance on Regulation of Genetically Engineered Animals" (article); "Animal Cloning: A Risk Assessment Draft Executive Summary" (report); *Animal Cloning and Food Safety* (brochure).

National Human Genome Research Institute (NHGRI)
URL: http://www.genome.gov/

The National Human Genome Research Institute was founded in 1989 as the National Center for Human Genome Research. Originally its function was to carry out the responsibilities of the National Institutes of Health in the Human Genome Project. As that project evolved, the mission of NHGRI changed, expanding in 1993, for example, to include a study of genetic diseases. In 1996, its mission was expanded further with the creation of the Center for Inherited Disease Research (CIDR) to study the genetic component of complex inherited disorders. In 1997, the center was given its present name and was elevated to a full institute within the National Institutes of Health (NIH), putting it on equal basis with 26 other institutes and centers that make up the NIH. Upon completion of its first mission, sequencing of the complete human genome in April 2003, the institute has moved on to other goals and objectives, including research on the ethical, legal, and social implications (ELSI) of genomic discoveries; developing strategies for using genomic information to reduce health disparities among Americans; and providing support for the training and education of the next generation of genomic researchers.

Publications: A very large number of books, reports, and articles, such as *Current Topics in Genome Analysis*, published annually, a lecture series on current research and development in genomics; *NHGRI Intramural Research Brochure*, summarizing the work of the Division of Intramural Research; *NIH Health Disparities Strategic Plan*, a publication of the Division of Extramural Research; reports from the Ethical, Legal, and Social Implications (ELSI) Research Program on topics such as genetic testing, genetic discrimination, issues related to the conduct of genetic research, specific research topics, and program reviews; and long-range planning reports and publications, such as *A Vision for the Future of Genomics Research* and reports from planning workshops and meetings.

National Institutes of Health (NIH)
See National Human Genome Research Institute; Office of Biotechnology Activities

Office of Biotechnology Activities (OBA)
URL: http://oba.od.nih.gov/oba/

The Office of Biotechnology Activities is an agency within the NIH with responsibility for promoting science, safety, and ethics in biotechnology by advancing knowledge, enhancement of public understanding, and development of sound public policies. Its activities include monitoring of scientific progress in human genetics research; coordinating and providing liaisons with federal and international organizations concerned with recombinant DNA research, human gene transfer, and genetic technologies; providing advice to the director of the NIH, other federal agencies, and state regulatory agencies with responsibilities for these fields of genetic research; developing and implementing policies associated with rDNA research and human gene therapy research and clinical studies; and responding to requests for information about recombinant DNA research and applications. The agency carries out its work through four major programs, the Recombinant DNA Advisory Committee (RAC), the Secretary's Advisory Committee on Genetics, Health, and Society (SACGHS), the National Science Advisory Board for Biosecurity (NSABB), and the Clinical Research Policy Analysis and Coordination (CRpac) Program.

Publications: A number of technical reports, administrative forms, reports of working groups, and educational documents, such as *Conclusions and Recommendations of the NIH Recombinant DNA Advisory Committee; Biosafety Considerations for Research with Lentiviral Vectors; Enhancing the Protection of Human Subjects in Gene Transfer Research at the National Institutes of Health; Human Gene Transfer Protocols;* and *Addressing Biosecurity Concerns Related to the Synthesis of Select Agents.*

Office of Public Health Genomics (OPHG)
URL: http://www.cdc.gov/genomics/

The Office of Public Health Genomics is a division of the Centers for Disease Control and Prevention (CDC) established in 1997 as the Office of Genetics and Disease Prevention, later renamed the Office of Genomics and Disease Prevention in 2003, and then given its present name in 2006. The agency's mission is to integrate the new field of genomics into public health research, policy, and programs in such a way as to improve the lives and health of all people. A topic of recent interest to OPGH has been the use of genetic testing in the diagnosis of diseases in newborn children, adults, and human fetuses.

Publications: *10 Years of Public Health Genomics at CDC 1997–2007* (report); *At A Glance—CDC's Office of Public Health Genomics—2008* (annual report); "Translating Genomics into Public Health Practice" (information kit); *Evaluation of Genomic Applications in Practice and Prevention (EGAPP): Implementation and Evaluation of a Model Approach* (report); "Genomics & Health Weekly Update" (online newsletter).

President's Council on Bioethics
URL: http://www.bioethics.gov/

The President's Council on Bioethics was created in 2001 by executive order of President George W. Bush. The council was renewed by new executive orders in 2003, 2005, and 2007. The council is the latest of a number of federal commissions appointed to consider bioethical issues, dating back to 1974, with the creation of the National Commission for the Protection of Human Subjects of Biomedical and Behavioral Research by the U.S. Congress. A history of earlier commissions and reports they produced is available at the council's Web site at http://www.bioethics.gov/reports/past_commissions/index.html. In its eight-year history, the council has considered a number of bioethical issues related to topics such as aging and end of life, biotechnology and public policy, cloning, death, genetics, health care, human dignity, nanotechnology, neuroethics, newborn screening, organ transplantation, research ethics, sex selection, and stem cells.

Publications: Many reports from this council and previous councils are listed on the council's Web site. Recent reports: *The Changing Moral Focus of Newborn Screening: An Ethical Analysis by the President's Council on Bioethics, Human Dignity and Bioethics: Essays Commissioned by the President's Council on Bioethics, Reproduction and Responsibility: The Regulation of New Biotechnologies,* and *Human Cloning and Human Dignity: An Ethical Inquiry.*

U.S. Environmental Protection Agency (EPA)
URL: http://www.epa.gov/

The U.S. Environmental Protection Agency is the primary agency for monitoring the environmental health of the United States. Among its many responsibilities is monitoring genetically engineered pesticides. Authority for this assignment rests in two important pieces of legislation, the Federal Insecticide, Fungicide, and Rodenticide Act (FIFRA), 7 U.S.C. §136, *et seq.*, and the Federal Food, Drug, and Cosmetics Act (FFDCA), 21 U.S.C. §301, *et seq.* With the development of methods for genetically engineering plants with the ability to produce pesticides and related products, responsibility for investigating and licensing these products fell to the EPA. Detailed information about the EPA's work in this regard is available on the agency's Web site under its section on Plant Incorporated Products (PIPs), at http://www.epa.gov/opp00001/biopesticides/pips/index.htm.

Publications: Copies of departmental regulations; case histories; relevant references from the *Federal Register*; fact sheets; reports.

United States Regulatory Agencies Unified Biotechnology Web Site
URL: http://usbiotechreg.nbii.gov/

In 1986, all federal agencies with responsibility for monitoring and regulating new biotechnology products combined to develop the Coordinated Framework for Regulation of Biotechnology, explaining in detail the specific responsibilities of each agency for products and methodologies developed in biotechnology. This Web site summarizes all information related to the Coordinated Framework, including a database of all actions taken on all engineered products used in agriculture in the United States, the role of each agency with regulatory authority in the field, laws and regulations related to the genetic modification of crops, government contacts for further information on regulatory issues, and links to additional sites related to this issue. A Web page on frequently asked questions provides a wealth of information on virtually every aspect of the regulation of genetically modified crops by the U.S. government.

Publications: None

Nongovernmental Organizations

Ag BioTech InfoNet
URL: http://www.biotech-info.net/

Ag BioTech InfoNet is a Web-based information system that deals with all applications of biotechnology and genetic engineering in the production, processing, and sale of food products. The major focus of the effort is on scientific and technical developments, although social, ethical, political, and economic issues are also reviewed. Information is arranged into two major categories: genetic engineering applications and impacts and implications. The former category includes topics such as insect resistance, herbicide tolerance, disease resistance, and other traits, along with special information for teachers and a list of resources and links. The latter category includes topics such as industry mergers and integration, rating biotech, applications, environmental impacts, health risks, costs and benefits, consumer choice, and policy.

Publications: None

AgBioWorld
URL: http://www.agbioworld.org/

AgBioWorld was founded in January 2000 by Professor C. S. Prakash of Tuskegee University and Gregory Conko of the Competitive Enterprise Institute with the goal of providing reliable scientific information on the use of biotechnology in agriculture to all concerned parties throughout the world. The organization is a 501(c)(3) nonprofit organization that depends primarily on volunteers for its operation. It attempts to remain rigorously nonbiased in its work, rejecting contributions of any organization with a vested interest in favor of or opposed to the use of biotechnology in agricultural operations. The major activity of AgBioWorld is providing information on agricultural biotechnology. In concert with that activity, the organization also provides referrals to experts in the field who can be contacted for additional, detailed information on a variety of topics, such as insect and herbicide tolerant crops, GM food safety and human health, feed safety and animal health, economic issues related to biotechnology, risk assessment and public perception of biotechnology, plant biology, breeding and development, food and trade policy, regulations and legal issues, and liability for GMOs and non-GMOs. AgBioWorld has also written a Declaration of Support for Agricultural Biotechnology that has now been signed by 3,400 scientists worldwide, including 25 Nobel Prize winners.

Publications: None

Agribusinness Accountability Initiative (AAI)
URL: http://www.agribusinessaccountability.org/bin/view.fpl/1194.html

The Agribusiness Accountability Initiative was founded in 2002 as a joint endeavor of the Center of Concern and the National Catholic Rural Life Conference. These two organizations continue to be responsible for different parts of the AAI project. AAI was created to challenge the evolving control of the world's food supply by a relatively small number of large corporations who, the project says, claims to work for the benefit of humans around the world but, in fact, have profit-making as their primary goal. AAI activities are centered on four themes: impacts, an explanation of the ways in which agribusiness controls the world's supply of foods; matrix, a collection of data on specific corporations involved in food production; global response, a review of efforts to combat the dominance of the food supply by large corporations; and clearinghouse, a collection of reports, data, information, and research on food companies and their impact on society. AAI provides a number of resources dealing with the application of biotechnology to the food supply.

Publications: None

American Board of Genetic Counseling (ABGC)
URL: http://abgc.iamonline.com/english/view.asp?x=1

The American Board of Genetic Counseling is the accrediting agency for genetic counseling in the United States and Canada. It establishes standards in the field by the accreditation of graduate programs in genetic counseling and conducting examinations that lead to diplomates as certified genetic counselors.

Publications: *American Board of Genetic Counseling* (brochure); fact sheets on licensure and recertification.

American Society of Bioethics and Humanities (ASBH)
URL: http://www.asbh.org/

The American Society of Bioethics and Humanities was formed in 1998 in the consolidation of three existing organizations, the Society for Health and Human Values, the Society for Bioethics Consultation, and the American Association of Bioethics. The mission of the organization is to promote the exchange of ideas and to foster multidisciplinary, interdisciplinary, and inter-professional scholarship, research, teaching, policy development, professional development, and collegiality among people engaged in clinical and academic bioethics and the medical humanities. ASBH conducts an annual meeting, a spring meeting, and an annual National Undergraduate Bioethics Conference, as well as endorsing and publicizing a number of other meetings by sister societies with interests and concerns compatible with its own. The organization also provides an online job bank for members.

Publications: *Publishing without Perishing* (handbook); *Core Competencies For Health Care Ethics Consultation* (handbook); *ASBH Reader* (journal).

American Society of Gene and Cell Therapy (ASGCT)
URL: http://www.asgt.org/

The American Society of Gene and Cell Therapy is a professional association of scientists and researchers working for the development of new technologies for gene and cell therapy. The organization was founded in 1996 with fewer than 1,000 members. It has grown to twice that size, with the largest number of members (about 1,400) from the United States. The association is interested in promoting research on new gene and cell therapies, providing opportunities for the exchange of information among researchers in the field, and promoting the development of methods by which research findings can be translated into clinical studies. In addition to its support for professional activities in gene and cell therapy, the society provides educational materials to the general public.

Publications: News releases and position statements; *Molecular Therapy* (journal).

American Society of Human Genetics (ASHG)
URL: http://www.ashg.org/

The American Society of Human Genetics was founded in 1948 as the primary organization of human geneticists worldwide. Today, its 8,000 members include researchers, academicians, clinicians, laboratory practice professionals, genetic counselors, nurses, and other individuals with a special interest in the field of human genetics. The organization has a fourfold list of objectives that includes sharing of recent research results through an annual convention and the association's professional journal; advocating for research support in the field of human genetics; providing educational materials for professionals and the general public; and promoting and supporting responsible scientific and social policies related to human genetics. The organization carries out its work through eight committees: awards, executive, finance, information and education, nominating, program, social issues, and professional development.

Publications: *American Journal of Human Genetics*; *SNP-IT Newsletter* (electronic newsletter); *Enhancement of K-12 Human Genetics Education: Creating a Cooperative Plan* (report); *Conversations in Genetics* (video recordings); *Medical School Curriculum Guidelines* (brochure).

Association of Forensic DNA Analysts and Administrators (AFDAA)
URL: http://www.afdaa.org

The Association of Forensic DNA Analysts and Administrators is a nonprofit organization of individuals involved in the forensic and legal applications of DNA typing. The organization currently has about 200 members from more than 75 government agencies and private corporations in 25 states. The organization meets about twice a year at the Texas Department of Public Safety in Austin. The objectives of the organization are to help members stay up to date with DNA technology, promote dissemination of information about DNA typing to the professional community, discuss legislative issues relating to forensic DNA typing, provide an opportunity for professionals in the field to network with each other, and offer formal training in DNA typing and analysis to members.

Publications: AFDAA Meeting Minutes (available online at the AFDAA Web site).

Biotechnology Industry Organization (BIO)
URL: http://bio.org/

The Biotechnology Industry Organization was formed in 1993 by the merger of the Association of Biotechnology Companies and the Industrial Biotechnology Association to provide a single voice for companies working in the field of biotechnology to speak about regulation of biotech crops, small business and economic development issues, national health care policy, and reform of the FDA. BIO claims a number of successes in the last 15 years of its existence, including enactment of reforms in the

FDA in the 1997 Food and Drug Administration Modernization Act, adoption of BIO principles on coverage of outpatient drug bills in the Medicare Modernization Act of 2003, dramatic growth in the amount of biotech crops planted each year, adoption by at least 40 states of programs to encourage biotech investment within the state, significant increases in farm income as a result of the use of biotech crops, and the cloning of endangered species using biotech technology.

Publications: *Guide to Biotechnology 2008* (report); *Technology, Talent, and Capital: State Bioscience Initiatives 2008* (report); *Biotechnology Solutions for Everyday Life* (brochure); *The Complete Connection* (brochure); *BIO News* (newsletter).

California Certified Organic Farmer (CCOF)
URL: http://www.ccof.org/

California Certified Organic Farmer (now known in almost all cases as CCOF) was founded in 1973 to provide a uniform, widely accepted system of certification for organic foods, one of the first such organizations created in North America. Today the organization consists of more than 1,600 food production and processing companies throughout Canada and the United States. Over the past four decades, CCOF has greatly expanded its activities and now sponsors conferences and other meetings, proposes and supports legislation, provides certification for organic products, supports educational programs and efforts, and organizes and coordinates political actions. Its efforts in the fields of genetic engineering and cloning have been focused on legislation monitoring and limiting genetically modified crops, support of local initiatives on genetic engineering, the suppl of information on GM crops to farmers and the general public, and prevention of the use of clones in agriculture.

Publications: *2008 Organic Directory*; *Certified Organic* (magazine); *CCOF E-Newsletter* (online newsletter); press releases, fact sheets, press kit, testimonials.

The Campaign to Label Genetically Engineered Foods ("The Campaign")

The Campaign to Label Genetically Engineered Foods (often known simply as "The Campaign") was founded in March 1999 as a 501(c)(4) nonprofit, non-tax deductible organization working to get legislation passed in the U.S. Congress that requires all foods containing genetically engineered components to be so labeled. The Campaign added other programs to its agenda, including one designed to have pharmed crops regulated and labeled, one to protect organic food from possible contamination with bioengineered products, and one to ban the production and sale of genetically modified rice. Craig Winters, founder of The Campaign, died in July 2009, and the organization is no longer active.

Publications: *Take Action Packet* (booklet); "Genetically Engineered Foods & Your Health"; "Genetically Engineered Foods & the Environment"; "Why Label Genetically Engineered Foods?"; "What Is Genetically Modified Food (And Why Should You Care)?" (flyers).

Center for Bioethics and Human Dignity (CBHD)
URL: http://www.cbhd.org/

The Center for Bioethics and Human Dignity was founded in 1993 by a group of Christian bioethicists who were concerned that Christian perspectives were not being included in discussions of a number of important socioscientific issues. These individuals joined together to form a nonprofit 501(c)(3) educational institution to remedy this problem, a program that is currently located at Trinity International University in Deerfield, Illinois. The center currently focuses its work on a number of topics in the field of bioethics, including biotechnology, cloning, end of life, genetics, health care and clinical ethics, health and spirituality, alternative medicine, neuroethics, reproductive ethics, sanctity of life, and stem cell research. Center researchers prepare position papers, bibliographies, and other research and educational materials on these topics, most available online at the center's Web site. The center has also sponsored conferences on topics in bioethics, including Global Bioethics, Healthcare and the Common Good, and Extending Life: Setting the Agenda for the Ethics of Aging, Death, and Immortality.

Publications: Many online resources.

Center for Food Safety (CFS)
URL: http://www.centerforfoodsafety.org/

The Center for Food Safety is a nonprofit advocacy membership organization founded in 1997 by the International Center for Technology Assessment. Its mission is to challenge potentially harmful food production technologies and to promote sustainable alternatives. The center uses a variety of methods to achieve its goal, including litigation, petitions for rule making, support for other agricultural and food safety organizations, grass roots organizing, public education, and media outreach. Its work is currently organized into 10 major campaigns: genetically engineered food, food and global warming, cloned animals, aquaculture, organic and beyond, national organic coalition, rbgh/hormones, food irradiation, sewage sludge, and mad cow disease.

Publications: *Your Right to Know: Genetic Engineering and the Secret Changes in Your Food* (book); *Food Safety Now!* (newsletter); *Food Safety Review*; many reports, including "Concentration in Cotton," "Contaminating the Wild," "A New View of U.S. Agriculture," "Food Irradiation: A Gross Failure," "California Food & Agriculture Report Card: Genetic Engineering."

Center for Genetics and Society (CGS)
URL: http://www.geneticsandsociety.org/

The Center for Genetics and Society was established in August 2001 as an outgrowth of an earlier group, the Exploratory Initiative on the New Human Genetic Technologies, a project of the Public Media Center in San Francisco. The center's position is that new genetic technologies have significant potential for improving the health and welfare of all human beings, but that they can also be used for less-than-salubrious purposes by researchers and industry. It works to educate the general public about the risks and benefits of new genetic technologies by sponsoring conferences and briefings, publishing reports, and providing support for other groups interested in these issues. The center focuses on two major topics: (1) "Biotechnology in the Public Interest," which includes the topic "Stem Cell Research: Ensuring Public Accountability" and (2) "Reframing the Human Biotech Debate: Beyond Embryo Politics, and Gender, Justice and Human Genetics," which includes the topics "Reproductive Health and Human Rights" and "Race, Disability and Eugenics."

Publications: Fact sheets, annual reports, op-eds and commentaries, talks and testimonies, articles; also, *Monthly Views and News* (newsletter); (report); *Playing the Gene Card? A Report on Race and Human Biotechnology* (report); *Next Steps for Stem Cell Research and Related Policies* (report).

Center for Genomics and Public Health (CGPH)
URL: http://depts.washington.edu/cgph/

The Center for Genomics and Public Health is affiliated with the School of Public Health and Community Medicine at the University of Washington, in Seattle. The mission of the center is to integrate new knowledge in genomics into public health practices. Its goal is to provide educational resources to state public health agencies and practitioners, insurance providers, and the general public and to develop ways for using genomics to improve the health of the general public. The center provides online training sessions and offers a summer institute on genetic testing and related issues.

Publications: "Gene Scene" (fact sheets); *Spotlight* (newsletter); journal articles published by center staff and researchers.

Clone Rights United Front
URL: http://www.clonerights.com/

Clone Rights United Front was founded in 1997 as an organization committed to advocacy on behalf of the right of human beings to be cloned. Its guiding philosophy is expressed in a Clone Bill of Rights which makes three assertions: (1) Every person owns his or her own DNA and has the right to use it for reproduction as he or she chooses; (2) That right is not assigned to any governmental, religious, or other organization, but is reserved to the person; and (3) Research and not rhetoric is the only legitimate way of determining the effects of cloning on human beings. The organization's Web site claims to hold more than 5,000 articles in its archives on the topic of human cloning.

Publications: None

Coalition for Genetic Fairness (CGF)
URL: http://www.geneticfairness.org/

The Coalition for Genetic Fairness was founded in 1997 to address issues regarding the use of genetic information in insurance and employment decisions. The organization originally consisted of civil rights, disease-specific, and health care organizations, but later expanded to include representatives of industry and employers. The coalition's mission has been to educate the general public and legislators about discrimination based on genetic testing and screening. That objective was accomplished in May 2008 when President George W. Bush signed the Genetic Information Nondiscrimination Act. Since the act has become law, the coalition has refocused its efforts on informing the health community and the general public on the provisions of the act and how they can best be implemented.

Publications: None

Council for Responsible Genetics (CRG)
URL: http://www.gene-watch.org/

The Council for Responsible Genetics is a nonprofit organization founded in Cambridge, Massachusetts, in 1983 for the purpose of providing a forum for discussions on the social, ethical, and environmental implications of genetic technologies. The council's activities are currently focused on eight major topics: genetic determinism, genetic testing, privacy and discrimination, cloning and human genetic manipulation, biotechnology and agriculture, biowarfare, genetic bill of rights, women and biotechnology. The council has produced a very large number of books, reports, brochures, position papers, online articles, and other resources on the whole range of topics in which it is interested.

Publications: *Gene Watch* (bimonthly publication); "Genetic Discrimination Position Paper"; "Position on Predictive Testing"; "Germline Manipulation Position Paper"; "Sexual Orientation Bibliography"; "Frequently Asked Questions About Genetically Engineered Food" (pamphlet).

DNA Academy
URL: http://www.albany.edu/nerfi/dna_academy/

The DNA Academy was established in 2004 at the University at Albany's Northeast Regional Forensics Institute. The motivation for the creation of the academy was concern about the slow pace of DNA analysis in the United States, reflecting in part a significant deficiency in the number of laboratory personnel trained in DNA technology. The academy currently offers a customized curriculum in DNA typing, a state-of-the-art DNA laboratory, highly qualified instructional staff, and a dramatically improved turnaround time in the handling of DNA specimens.

Publications: *2007 AAFS Presentation* (power point presentation at the 2007 American Academy of Forensic Sciences conference); *Basic DNA Academy Curriculum* (academy catalog of course offerings).

European NGO Network on Genetic Engineering (GENET)
URL: http://www.genet-info.org/

The European NGO Network on Genetic Engineering is an international nonprofit organization chartered under Swiss law. It was founded in 1995 for the purpose of providing information on genetically modified organisms to member organizations and individual citizens and to support activities and campaigns against genetic engineering. As of 2009, 51 organizations in 27 countries were members of GENET. The organization focuses on the distribution of information on plant and animal breeding, human health, agriculture, animal welfare, and food production as they relate to biological diversity, human genetics and medicine, the environment, and socioeconomic development. GENET sponsors two meetings annually, one an annual conference and the second a member meeting or public conference on some specific topic. The primary mechanism by which GENET achieves its objectives are three online mailing lists: GENET-news, GENET-forum, and GENET-action. GENET-news provides access to journal articles, press releases, Internet documents, and member messages on a variety of topics such as: production and use of recombinant proteins; breeding and use of transgenic plants, their ecological and biodiversity risks; production and use of food made from or produced by GMOs; production and use of enzymes and other food ingredients synthesized by GMOs; patenting of genes and transgenic organisms; biopiracy; breeding and use of transgenic animals in agriculture and "gene pharming"; xenotransplantation; national and international laws covering genetic engineering; economics and company news; and human genome projects, human cloning, and human gene therapy. GENET-forum is a mailing list for strategy and political discussions among GENET members. GENET-action is a mailing list for members to organize joint activities and coordinate actions.

Publications: "GENET—European NGO Network on Genetic Engineering" (information sheet).

GE Food Alert
URL: http://www.gefoodalert.org/

GE Food Alert was founded in 2000 by a coalition of seven groups opposed to the production and sale of genetically modified foods. The organization was involved in a number of efforts to block the use of GM foods in consumer products, most famously in its battle against the StarLink variety of genetically modified corn in the early 2000s. GE Food Alert no longer exists as an activist organization; most of its efforts are now included in the mission of the Institute for Agriculture and Trade Policy (IATP), one of the organizations which established GE Food Alert in the first place. GE Food Alert continues to exist, however, as a repository of information about the risks posed by genetically modified foods and about efforts to combat the spread of these products. More detailed information about IATP and its policies and programs is available at its Web site, http://www.iatp.org/.

Publications: None

Genetic Alliance
URL: http://geneticalliance.org/

Genetic Alliance is a 501(c)(3) nonprofit organization that seeks to promote the use of new genetic technologies in the treatment of human diseases. It is a collaborative of community, governmental, industrial, and private agencies with an interest in this issue. Three programs through which it carries out its mission are: (1) Access to Credible Genetics Resources Network, a cooperative program with the Centers for Disease Control and Prevention to improve public access to new information about genetic technologies; (2) Consumer Focused Newborn Screening, an effort to find ways of maximizing the benefits, while minimizing the risks, of newborn genetic screening protocols; and (3) Community Centered Family Health History, designed to improve the use of family health histories for making current health decisions.

Publications: monographs, a policy bulletin, a weekly bulletin, how-to guides; *Guide to Understanding Genetics for Patients and Professionals* (book; available online); *Making Sense of Your Genes: A Guide to Genetic Counseling* (book; available online); *Genetic Testing* (journal of Genetic Alliance).

Genetic Engineering Action Network (GEAN)
URL: http://www.geaction.org/

Genetic Engineering Action Network is a consortium of about 100 national, state, regional, and local organizations opposed to the use of genetic engineering in agriculture. The network sponsors and cooperates with a wide variety of efforts to achieve this objective, ranging from the passage of local laws against genetically engineered crops to the adoption of national and international laws, policies, and regulations that limit or prevent the use of genetic engineering in the development of crops. The group's four core principles are: (1) a demand for labeling of genetically engineered foods and feeds so that consumers may know when they are purchasing GM foods; (2) assertive programs of control and management of genetically engineered crops to make sure that they meet all applicable standards; (3) adoption and enforcement of antitrust legislation to prevent large corporations from making it difficult for small farmers to survive and succeed without making use of genetically engineered crops; and (4) establishment of clear systems to ensure that corporations are responsible for liabilities resulting from accidents or other harmful effects of bioengineered crops.

Publications: *Local Organizing Toolkit* (online instructional manual).

Genetic Engineering Network (GEN)
URL: http://www.geneticsaction.org.uk/

Genetic Engineering Network is a British organization committed to working against all forms of genetic engineering, including the production of genetically modified foods and human and animal genetic engineering, and against biotech corporations who use and control these technologies. The network consists of a number of other organizations, many of which are concerned with one or more specific aspect of these issues. These organizations include Corporate Watch, Earth First!, genetiX snowball, Genetix Food Alert, Five Year Freeze, Green Party, Greenpeace, Women's Environmental Network and Gaia Foundation, as well as retailers in the natural food trade, organic farmers, and local groups and individuals who are opposed to the use of genetic engineering in agriculture. GEN coordinates and contributes to the activities of these individual groups as well as provides information about developments in the field of genetic engineering.

Publications: *GeneTix* (newsletter).

Genetic Engineering Policy Alliance (GEPA)
URL: http://www.gepolicyalliance.org/

The Genetic Engineering Policy Alliance is a coalition of organizations and individuals promoting precautionary policies and legislation with regard to the use of genetically engineered materials in agriculture. Its work is focused on three major concerns: (1) protecting farmers from economic damage as a result of the accidental dispersion of genetically engineered seeds and crops; (2) promoting the public's right to know about the presence of genetically engineered products in its food supplies; and (3) safeguarding the environment and human health from possible deleterious effects resulting from exposure to genetically engineered products. The organization's general philosophy is outlined in its Policy Platform, online at http://www.gepolicy alliance.org/policy_platform.htm, which members are required to acknowledge as a condition of joining the coalition. GEPA currently has about 70 members including a wide array of groups, such as the Agricultural Land-Based Training Association, Food and Water Watch, Straus Family Creamery, Frey Vineyards, Whole Foods Market, Californians for Alternatives to Toxics, Occidental Arts and Ecology Center, Episcopal Diocese of California, and Ella Baker Center. GEPA's Web site contains a large collection of articles related to the use of genetic engineering in agriculture.

Publications: *Genetic Engineering Policy Alliance Newsletter* (online newsletter).

Greenpeace
URL: http://www.greenpeace.org/international/; http://www.greenpeace.org/usa/

Greenpeace was founded in 1971 when a group of environmentalists leased a small fishing vessel to protest nuclear tests being conducted off the coast of Alaska. Since that time, the organization has grown to become one of the largest environmental groups in the world with national chapters in the United States and more than 40 other countries. The organization's mission has also expanded to include a number of environmental issues, including global climate change, the oceans, ancient forests, peace and disarmament, nuclear weapons, genetic engineering, toxic chemicals, and sustainable trade. Among the many victories claimed by the organization are: a successful effort to reduce the destruction of forests in Indonesia for the construction of palm oil plantations; passage of a ban on incandescent light bulbs in Argentina; adoption of a ban on deep-sea ocean bottom trawling off the coast of Chile; a successful public vote on a ban on genetically engineered crops in Switzerland; discontinuation of all genetically engineered crops in India by the Bayer Corporation; and suspension of tests on genetically engineered Round Up by the Monsanto corporation.

Publications: A very large number of reports, including *Greenpeace Criteria for Sustainable Fisheries*; *Cattle Ranching Expansion in the Brazilian Amazonia*; *GE Trees Briefing*; *Monsanto's 7 Deadly Sins*; *GM Contamination Register Report*.

Innocence Project
URL: http://www.innocenceproject.org

The Innocence Project was established in 1992 at the Benjamin N. Cardozo School of Law of Yeshiva University in New York City by Barry C. Scheck and Peter J. Neufeld. Both men had been involved in cases involving the use of DNA typing from its earliest days. The purpose of the project was to assist individuals who had been improperly convicted of crimes. As of early 2009, 234 individuals had been exonerated of the crimes for which they had been convicted, based on DNA evidence. In addition to working with individual prisoners who claim they have been wrongfully convicted, the project works for the passage of enlightened legislation at the state and federal levels to make possible the exoneration of innocent victims of the law.

Publications: newsletters, fact sheets, and press releases describing the project's work and accomplishments; an online blog.

Institute for Responsible Technology (IRT)
URL: http://www.responsibletechnology.org/GMFree/Home/index.cfm

The Institute for Responsible Technology was founded in 2003 by Jeffrey Smith, a former marketing consultant who became interested in the risks posed by genetically modified foods in the human diet. The mission of IRT is to educate the general public about the risks of eating GM foods and about healthier alternatives to engineered foods. IRT's efforts to exclude GM foods from the American diet are expressed in its Campaign for Healthier Eating in America, whose goal is to stop the use of genetic engineering in the American food industry. The institute is also involved in the development of a GM-free schools movement, which focuses its non-GMO foods efforts on the foods served in American schools. IRT's Web site is a valuable resource for information on: the fundamentals of genetically modified organisms and foods, the use of recombinant bovine growth hormone (rBGH) in dairy products, ways of shopping for non-GMO foods, and opportunities for activism against engineered foods.

Publications: "The Non-GMO Shopping Guide" (brochure); "Health Risks Brochure"; *Genetic Roulette: The Documented Health Risks of Genetically Engineered Foods* (book by Jeffrey Smith); *Seeds of Deception* (book by Jeffrey Smith); *Spilling the Beans* (e-newsletter); many DVDs and CDs, including *The GMO Trilogy, Hidden Dangers in Kids' Meals,* and *You're Eating What?*

International Biopharmaceutical Association (IBPA)
URL: http://www.ibpassociation.org/index.php

The International Biopharmaceutical Association was founded in 1990 to bring together institutions and corporations from nations around the world to work on issues of common concern. The association collaborates with international, regional, and national bodies concerned with the biopharmaceutical industry and clinical research. Its services are available to organizations, institutions, regulatory authorities, individual policy and decision-makers, specialists, administrators, researchers, educators, and students. As of 2009, IBPA had more than 7,000 individual and 200 corporate members from every part of the world. The association takes part in a number of meetings, conferences, and events at companies, academic institutions, and other institutions every year. A major part of IBPA's focus is career counseling. It provides information on careers in biopharmaceuticals, lists job openings, offers assistance with resume writing and job searching, provides coaching for careers, and provides information on internships.

Publications: *IBPA Newsletter; Proceedings of the International Biopharmaceutical Association* (journal); *BioAuditor* (journal).

National Society of Genetics Counselors (NSGC)
URL: http://www.nsgc.org/

The National Society of Genetics Counselors was founded in 1979 to be the primary advocacy group for genetics counselors and to ensure the availability of quality genetics counseling through education, research, and the promotion of public policy in the field. Important elements of the organization's work are carried out through an annual educational conference, occasional regional conferences, an online educational conference, and continuing education courses. The organization's Web site provides information and resources for health care professionals and for the general public. It also explains the process by which one becomes a genetics counselor, with referrals to programs available in the field.

Publications: *Perspectives in Genetic Counseling* (newsletter); *Journal of Genetic Counseling*; also, many brochures, pamphlets, handbooks, and other references on specific diseases and procedures, such as *Genetic Testing for CF: A Handbook for Families; Genetic Testing for CF: A Handbook for Professionals; Testing for Huntington's Disease: Making an Informed Choice* (booklet); *An Ethics Casebook for Genetic Counselors, 2nd Edition: Ethical Discourse for the Practice of Genetic Counseling; Cancer Risk Genetic Counselors: Working Hand in Hand with Physicians* (brochure).

Non-GMO Project
URL: http://www.nongmoproject.org/

The Non-GMO Project originated at a single grocery store, The Natural Grocery Company, in Berkeley, California, in 2003. It was created because of concerns by customers that they were uncertain as to the presence of genetically modified foods at Natural Grocery. The owners of Natural Grocery decided that they needed clear standards for defining genetically modified foods and an independent, disinterested source for verifying the status of GM and non-GM foods. They were eventually joined in this effort by The Big Carrot Natural Food Market in Toronto, Ontario, which had been working for well over a year to develop an identification and labeling process. In 2007, Non-GMO had expanded to include all stakeholder groups in the natural products industry, including consumers, retailers, farmers, and manufacturers. The organization also began to form a number of advisory boards on a number of technical and policy issues. The process by which food products can be verified as non-GMO and listed as approved by the Non-GMO Project, as well as a list of participants in the program, are available on the organization's Web site.

Publications: *Non-GMO Project Working Standard* (guideline).

Northwest Resistance against Genetic Engineering (NW RAGE)
URL: http://www.nwrage.org/

Northwest Resistance against Genetic Engineering describes itself as "a non-violent, grassroots organization dedicated to promoting the responsible, sustainable and just use of agriculture and science . . . [that is] working towards a ban on genetic engineering and patents on life." Some issues on which NW RAGE is currently working are agrofuels, genetically engineered trees, recombinant bovine growth hormone (rBGH), Roundup Ready creeping bent grass, and safe foods. Some specific goals of the organization are a ban on genetic engineering, a ban on patents on any kind of life form, a ban on biopiracy (defined as the theft of indigenous people's genes and knowledge), a ban on cloning of humans and animals, a rescinding of all current FDA approvals for genetically engineered products on the market, and the cessation of factory farming. The organization's Web site is a rich resource of shopping resources, industrial contacts, films and podcasts, downloads, and local resources. It also provides access to hundreds of articles dealing with the human health and environmental aspects of genetic engineering.

Publications: None

Organic Consumers Association (OCA)
URL: http://www.organicconsumers.org/

The Organic Consumers Association was formed in 1998 in reaction to proposed regulations for organic foods proposed by the USDA. The organization has continued to work for the creation and enforcement of strict standards for the growing and labeling of organic foods. It claims to be the sole national organization working on behalf of "the nation's estimated 50 million organic and socially responsible consumers." OCA's political efforts are guided by a six-point platform adopted in 2005 that outlines the organization's goals and objectives. OCA maintains a news section page on its Web site that contains print and electronic articles on topics related to genetically engineered foods.

Publications: "Hazards of Genetically Engineered Foods and Crops: Why We Need A Global Moratorium" (fact sheet in English and Spanish).

Pew Charitable Trusts
URL: http://www.pewtrusts.org/

Pew Charitable Trusts is an independent, nonprofit organization founded as the result of seven gifts between 1948 and 1979 from two sons and two daughters of Joseph N. Pew, founder of the Sun Oil Company, and his wife Mary Anderson Pew. The trusts currently support more than 100 distinct programs, ranging from youth voting to student debt to gay marriage to economic mobility to death penalty reform. In the area of science, it focuses on five subjects: agricultural biotechnology, biomedical research, environmental science, genetics and public policy, and nanotechnology. For each of these topics, Pew usually produces one or more reports on the subject, press releases and speeches dealing with the reports, fact sheets, and summary papers.

Publications: *U.S. vs. EU: An Examination of the Trade Issues Surrounding Genetically Modified Food* (report); *State Legislatures Continue to Be Active in Addressing Challenges Associated With Agricultural Biotechnology* (report); *The Genetic Town Hall: Public Opinion About Research on Genes, Environment, and Health* (report); *Public Health at Risk: Failures in Oversight of Genetic Testing Laboratories* (report); *Cloning: A Policy Analysis* (report); *Human Germline Genetic Modification: Issues and Options for Policymakers* (report).

Sierra Club
URL: http://www.sierraclub.org/

The Sierra Club is the oldest and largest grass-roots environmental organization in the United States. It was founded in 1892 by naturalist John Muir with 182 members. It was successful in its first environmental campaign, opposing an effort by the U.S. government to reduce the size of Yosemite National Park. Over the years, it has been at the forefront of most environmental controversies in the United States, although its focus has continued to expand and shift with changing conditions in the nation. Currently the six programs it emphasizes most strongly are "Curbing Carbon," "Move Beyond Coal," "Clean Energy Solutions," "Green Transportation," "Resilient Habitats," and "Safeguarding Communities." It continues to advocate for a number of other causes, however, including clean air, clean water, cool cities, environmental education, environmental justice, environmental law, global population, responsible trade, toxics, and wild lands. Its Web page on genetic engineering provides information on a number of topics, including the use of recombinant bovine growth hormone (rBGH) in dairy products, genetically engineered trees, genetically modified foods in school meals, privatization of wealth as a result of genetic engineering, and genetically engineered wheat.

Publications: more than 100 books, articles, and other materials covering a range of topics; *Hey, Mr. Green* (book); *The Case against the Global Economy* (book); *Coming Clean* (book); *Nature's Operating Instructions* (book); *The Unsettling of America* (book); *Sierra* (magazine).

Union of Concerned Scientists (UCS)
URL: http://www.ucsusa.org/

The Union of Concerned Scientists was formed in 1969 as a collaborative effort between students and faculty members at the Massachusetts Institute of Technology (MIT). Its earliest campaigns were against nuclear power plants and certain types of weapon systems being developed by the U.S. military. In recent years, it has greatly expanded its field of interests, now operating programs in global climate change, clean energy, clean transportation, nuclear power, nuclear weapons, food and agriculture, and invasive species. Its efforts in the area of food and agriculture involve the promotion of sustainable agriculture methods and improved government regulation of genetically engineered organisms and products. UCS lists its recent successes in this field as including the shaping of legislation supporting organic and sustainable agriculture, achieving a meaningful label for grass-fed meat, preventing a type of human medicine from being used in animal agriculture, pressing for a ban on the production of drugs and industrial chemicals in food crops, and strengthening oversight of pharma crops and cloning.

Publications: reports on a wide variety of topics, such as *The Consumer's Guide to Effective Environmental Choices: Practical Advice from the Union of Concerned Scientists; How to Avoid Dangerous Climate Change; Nuclear Power in a Warming World; Greener Pastures: How Grass-fed Beef and Milk Contribute to Healthy Eating; The Economics of Pharmaceutical Crops: Potential Benefits and Risks for Farmers and Rural Communities; A Growing Concern: Protecting the Food Supply in an Era of Pharmaceutical and Industrial Crops;* and *Gone to Seed: Transgenic Contaminants in the Traditional Seed Supply.*

8

Resources

The use of DNA technology for a variety of applications, including forensic science, genetic testing, human gene therapy, development of foods and related projects through genetic modification methods, pharming, and cloning is the subject of a very large number of books, articles, scientific and other reports, and Internet Web pages. In addition, a number of resources on the history and background of DNA technology are available in both print and electronic form. This chapter provides a sampling of resources currently available on these topics. Although the chapter is divided into separate sections by topic and by format, some resources may overlap these somewhat artificial distinctions.

History and Background

Books

Clark, David P., and Nanette Jean Pazdernik. *Biotechnology: Applying the Genetic Revolution.* Amsterdam; Boston: Academic Press/Elsevier, 2009.

This book provides a broad, general introduction to the subject of biotechnology and its many applications. Specific chapters deal with topics such as the basics of biotechnology; DNA, RNA, and protein; genomics and gene expression; protein engineering; transgenic plants and plant biotechnology; transgenic animals; gene therapy; biowarfare and bioterrorism; and ethical issues in the use of biotechnology.

Dale, Jeremy, and Malcolm von Schantz. *From Genes to Genomes: Concepts and Applications of DNA Technology.* Chichester, England; Hoboken, NJ: John Wiley & Sons, 2007.

This book covers virtually every aspect of DNA technology, from an explanation of the molecular structure of DNA to the techniques used in DNA technology to the applications of DNA technology in today's world.

Hodge, Russ. *Genetic Engineering: Manipulating the Mechanisms of Life.* New York: Facts On File, 2009.

This book provides a general overview of the history, technology, and ethical issues involved in DNA technology. Various chapters focus on the principles of classical genetics and molecular genetics, the rise of recombinant DNA technology, some practical applications of rDNA technology, and ethical issues associated with genetic engineering.

Jerslid, Paul T. *The Nature of Our Humanity: Ethical Issues in Genetics and Biotechnology.* Minneapolis, MN: Fortress Press, 2009.

The author is professor emeritus of theology and ethics at Lutheran Theological Southern Seminary and outlines the theological and ethical issues that arise out of biotechnological research on humans in the field of medicine and scientific research.

McCabe, Lina L., and Edward R. B. McCabe. *DNA: Promise and Peril.* Berkeley: University of California Press, 2008.

This book presents a number of issues that have arisen and may arise from scientists' ability to manipulate the DNA structure of individual organisms. Topics include the use of DNA technology in forensic science, protection against genetic discrimination, the ownership of genes by corporations, reproductive technologies, therapeutic and reproductive cloning, genomic medicine, and gene therapy.

Smith, Gina. *The Genomics Age: How DNA Technology Is Transforming the Way We Live and Who We Are.* New York: AMACOM (American Management Association), 2005.

Intended for a general audience with no specialized training in biotechnology, this book reviews the scientific basis for DNA technology and discusses the many applications it has in everyday life.

Thieman, William J., and Michael Angelo Palladino. *Introduction to Biotechnology*, 2nd ed. San Francisco: Pearson/Benjamin Cummings, 2009.

This book has been a standard textbook in the field for many years. It provides a broad, somewhat technical background to many aspects of the field of biotechnology, including an introduction to genes and the genome, the use of microbes in biotechnology, agricultural biotechnology, medical biotechnology, and ethical and legal issues.

Articles

Jeffreys, Alec J., Victoria Wilson, and Swee Lay Thein. "Hypervariable 'Minisatellite' Regions in Human DNA." *Nature* 314 (March 7, 1985): 67–73.

Although very technical, this article is of historical importance as the first scholarly article describing the process of DNA testing as a means of identifying individual humans.

Krimsky, Sheldon. "From Asilomar to Industrial Biotechnology: Risks, Reductionism and Regulation." *Science as Culture.* 14 (4; December 2005): 309–323.

Recombinant DNA research has always been beleaguered by ethical, social, economic, political, and legal issues. This essay reviews the evolution of those issues beginning with the Asilomar Conference held in February 1975 when the risks of rDNA were first discussed and acted upon by a group of scientists in the field.

Miller, Henry I., and Gregory Conko. "NGO War on Biotechnology." *Journal of Commercial Biotechnology.* 11 (3; 2005): 209–222.

The authors argue that critics of biotechnology incorrectly suggest that the risks of recombinant DNA technology differs significantly from historical efforts to manipulate the genome of plants and animals and that they make unrealistic claims for the need for control and regulation of such research.

Moran, Enda, and Richard Fink. "Biosafety for Large-scale Operations Using Recombinant DNA Technology: No Emerging Hazards." *Biopharm International.* **18 (9; 2005): 32–41.**

The authors were involved in the startup of a new biopharmaceutical plant in Ireland and were concerned about possible safety risks in the operation of the plant. They conducted a detailed review of the history of safety regulations and procedures for recombinant DNA research and production. This paper summarizes the results of their research on this topic.

Yi, Doogab. "Cancer, Viruses, and Mass Migration: Paul Berg's Venture into Eukaryotic Biology and the Advent of Recombinant DNA Research and Technology, 1967–1980." *Journal of the History of Biology.* **41 (4; December 2008): 589–636.**

This is an interesting survey of the early history of recombinant DNA research and a significant change in the paradigm of that research based on the work of Paul Berg. The latter portions of the article discuss some early and more recent applications of recombinant DNA research.

Reports

Office of Science and Technology Policy. "Coordinated Framework for Regulation of Biotechnology." *Federal Register* **51 (June 26, 1986): 23302.**

This document is of primary importance because it sets out U.S. policy on the regulation of all forms of biotechnology, explaining which agencies are responsible for which fields of research and laying out the policies and regulations of each agency with regulatory responsibility. This document is available online at http://usbiotechreg.nbii.gov/CoordinatedFrameworkForRegulationOfBiotechnology1986.pdf.

Internet Sources

"Chapter 8 A. Recombinant DNA Technology." URL: http://biology.kenyon.edu/courses/biol114/Chap08/Chapter_08a.html . Accessed on March 11, 2009.

This Web page was written for a college-level course in biology. It provides a moderately advanced explanation of the process by which recombinant DNA experiments are conducted and a good general introduction to the history and possible applications of the technology. Colorful graphics add to the quality of the presentation.

Fuller-Espie, Sherry. "Applications of Recombinant DNA Technology." URL: http://www.cabrini.edu/sepchedna/2002% 20 Workshop / Power % 20 Point % 20 Presentations / Lecture _ 3 _DNA_Workshop%C2%AD_NO_FIGS.ppt. Accessed on March 11, 2009.

This Web site offers a clear, well-designed Power Point presentation on recombinant DNA technology that explains the procedures used and some major applications of the technology in a variety of fields. This is an excellent introduction to the topic.

Lei, Hsien-Hsien. "Top 10 Ways DNA Technology Will Change Your Life." URL: http://healthnex.typepad.com/web_log/2007/ 05/top_10_ways_dna.html. Accessed on March 11, 2009.

This Web site is of particular value because it provides links to a number of specific examples of the application of recombinant DNA technology in a variety of fields.

"Recombinant DNA and Gene Cloning." URL: http://users .rcn.com/jkimball.ma.ultranet/BiologyPages/R/Recombinant DNA.html. Accessed on March 12, 2009.

This Web site provides an excellent, succinct, well-presented general introduction to the process by which recombinant DNA procedures are carried out. The presentation is technical, but at a level that most individuals should be able to understand.

University of Delaware. "rDNA." URL: http://present.smith .udel.edu/biotech/rDNA.html. Accessed on March 12, 2009.

This instructional program offers an animated or print description of the steps by which recombinant DNA technology is completed. The presentation is clear and at a technical level available to most readers.

Forensic Science

Books

Buckleton, John S., Christopher M Triggs, and Simon J Walsh. *Forensic DNA Evidence Interpretation*. Boca Raton, FL: CRC Press, 2005.

This book begins with a review of the chemistry and biology of DNA typing and then focuses on the statistical interpretation of data obtained from the analysis of DNA typing. The presentation tends to be quite technical, but it is of value even to the general reader.

Kobilinsk, Lawrence F., Thomas F Liotti, and Jamel Oeser-Sweat. *DNA: Forensic and Legal Applications*. Hoboken, NJ: Wiley-Interscience, 2005.

This book discusses virtually every aspect of DNA forensic typing, from the collection of evidence through its analysis to the presentation of data and interpretation in criminal cases

Lazer, David, ed. *DNA and the Criminal Justice System: The Technology of Justice*. Cambridge, MA: MIT Press, 2007.

The 16 chapters in this book deal with a variety of questions relating to the use of DNA typing in the criminal justice system, including current and future trends in the field, lessons from the history of fingerprinting for the use of DNA typing, genetic privacy, DNA databases, and selective arrests and prosecutions based on DNA information.

Newton, David E. *DNA Evidence and Forensic Science*. New York: Facts On File, 2008. A book in the Library in a Book series that provides an extended introduction to the topic, along with a variety of resources to aid the average researcher, including a group of biographical sketches, a chronology, list of organizations, list of print and electronic resources, and glossary of important terms.

Rudin, Norah, and Keith Inman. *An Introduction to Forensic DNA Analysis*, 2nd ed. Boca Raton, FL: CRC Press, 2002.

A standard textbook in the field of DNA typing, this work discusses the topic in great detail and at a level that is understandable to most well-educated individuals.

Articles

Aronson, Jay D. "Creating the Network and the Actors: The FBI's Role in the Standardization of Forensic DNA Profiling." *Biosocieties* 3 (2008):195–215.

The author provides an interesting historical review of the early development of DNA databases, issues raised about the development of this technology, and solutions developed by the FBI for dealing with the new technology of DNA fingerprinting.

Choi, H.-J., M. Ko, and J. H. Ahn. DNA Fingerprinting Using PCR: A Practical Forensic Science Activity Based on Inquiry-based Problem Solving." *Journal of Biological Education*. 43 (1: Winter 2008): 36–40.

The authors describe an educational program developed for college students, who are asked to find a putative suspect using DNA fingerprinting using a drop of blood and a hair root. Students were generally successful in carrying out the assignment and identifying the correct suspect.

"DNA Databases." *CQ Researcher*. 9 (20; May 28, 1999), whole.

This issue of the journal is devoted to a review of the status of DNA databases on the federal and state levels in the United States, and the ethical and legal issues raised by such databases. Included is a chronology of the use of DNA typical in forensic science, a review of some important cases, and some predictions for the future of DNA databases in this country.

"Inside the Mind of a Juror—The Problem with DNA: Forensic Evidence Increasingly Includes Genetic Fingerprinting, but Researchers Worry That Juries Put Too Much Stock in the Results. *Monitor on Psychology*. 38 (6; June 2007): 52.

This article discusses the problem jurors face in having to assess DNA typing information presented to them during a criminal case.

M'charek, Amade. "Contrasts and Comparisons: Three Practices of Forensic Investigation." *Comparative Sociology.* 7 (3, 2008): 387–412.

The author explains how DNA typing has created some theoretical issues about the "population" of individuals to whom a suspect may belong, but notes that forensic science has moved forward in the use of the technology without being concerned about the theoretical problems. He then reviews the three ways in which DNA typing is currently used to identify individuals involved in criminal activity.

Spagnoli, Linda. "Beyond CODIS: The Changing Face of Forensic DNA Analysis."*Law Enforcement Technology.* 34 (7; July 2007): 42, 44, 51.

The FBI's Combined DNA Index System (CODIS) has been very successful in collecting and collating DNA data collected from a variety of sources, but it has had limited success in identifying specific individuals who may have been involved in a crime. This article explains how changes in DNA technology are helping to resolve that problem.

Reports

Connors, Edward, Thomas Lundregan, Neal Miller, and Tom McEwen. *Convicted by Juries, Exonerated by Science: Case Studies in the Use of DNA Evidence to Establish Innocence After Trial.* Washington, D.C.: Office of Justice Programs. U.S. Department of Justice, June 1996.

This report reviews the use of DNA typing in the United States to exonerate men and women convicted of crimes, usually rape, aggravated assault, and/or murder. The report offers a number of recommendations to increase the efficacy of DNA typing in proving the innocence of prisoners and concludes with detailed reviews of more than two dozen individuals whose sentences were commuted as a result of DNA testing.

National Commission on the Future of DNA Evidence. *Postconviction DNA Testing: Recommendations for Handling Requests.* Washington, D.C.: U.S. Department of Justice. Office of Justice Programs. National Institute of Justice, 1999.

One of the special issues considered by the National Commission on the Future of DNA Evidence (see reference above) was the use of DNA typing results for the exoneration of wrongfully-convicted individuals. This report summarizes research done by the commission on this issue and its recommendations on the use of DNA evidence in such cases.

National Commission on the Future of DNA Evidence. *The Future of Forensic DNA Testing.* Washington, D.C.: U.S. Department of Justice. Office of Justice Programs. National Institute of Justice, 2000.

This report was written by a committee appointed by then-Attorney General Janet Reno to consider future applications of DNA typing in the United States. The report consists of two parts, the first of which is a general introduction to the subject and a consideration of possible future applications of DNA typing written for the general public. The second part of the report consists of a series of appendices that deal with technical issues involved in the use of DNA typing in forensic science.

National Commission on the Future of DNA Evidence. *What Every Law Enforcement Officer Should Know about DNA Evidence: A Pocket Guide.* Washington, D.C.: U.S. Department of Justice. Office of Justice Programs. National Institute of Justice, 1999.

This short pamphlet (described as a "report" by the commission) provides basic information about DNA typing, circumstances in which it can and should be used in law enforcement activities, and special cautions to observe in its use.

National Institute of Justice. *Report to the Attorney General on Delays in Forensic DNA Analysis.* Washington, D.C.: U.S. Department of Justice. Office of Justice Programs, March 2003.

One of the most difficult problems facing the use of DNA evidence in forensic science is the length of time required to obtain analyses of evidence collected at a crime scene. Wait times of many months are not unusual for prosecutors and other law enforcement officers hoping to use such data in crime investigations and prosecutions. This report analyzes the current status of delays in DNA testing and interpretation and changes needed in the future to make this technology more useful to the criminal justice system.

Nuffield Council on Bioethics. *The Forensic Use of Bioinformation: Ethical Issues.* London: Nuffield Council on Bioethics, September 2007.

This report addresses a number of current issues related to the use of fingerprints, DNA evidence, and other types of bioinformation by law enforcement agencies in the United Kingdom. It deals with issues such as the circumstances under which police officers should be permitted to take digital and DNA fingerprints, how long those prints and the profiles made from them should be retained, how bioinformation should and should not be used in criminal trials, and what the proper role of a national DNA database is.

Roman, John K., et al. *The DNA Field Experiment: Cost-Effectiveness Analysis of the Use of DNA in the Investigation of High-Volume Crimes.* Washington, D.C.: Urban Institute. Justice Policy Center, March 2008.

This report deals with the issue of DNA testing in so-called high-volume crimes, such as burglary, for which DNA typing has historically not been applied. The report summarizes evidence obtained from research conducted in five urban areas, Phoenix, Arizona; Orange County, California; Denver, Colorado; Los Angeles, California; and Topeka, Kansas. It finds substantial reasons for making use of DNA typing in "minor" crimes as well as in murder, aggravated assault, rape, and other so-called "major" crimes.

U.S. Department of Justice. *Advancing Justice through DNA Technology.* Washington, D.C.: Government Printing Office, March 11, 2003.

This report outlines an initiative of the same name proposed by President George W. Bush that promotes the use of DNA typing in solving crimes. The full report is available online at http://www.usdoj.gov/ag/dnapolicybooktoc.htm.

Internet Sources

DNA Initiative. "History of Forensic DNA Analysis." URL: http://www.dna.gov/basics/analysishistory/. Accessed on March 11, 2009.

This Web page is one of the best sources for unbiased information on the use of DNA typing in forensic science. It contains separate sections on the history of forensic DNA technology, basic biological concepts, techniques for and issues related to the gathering of DNA evidence, and the analysis and interpretation of DNA typing results.

Federal Bureau of Investigation. "CODIS." URL: http://www.fbi.gov/hq/lab/html/codis1.htm. Accessed on March 11, 2009.

CODIS is the Combined DNA Index System operated by the U.S. Federal Bureau of Investigation (FBI). This Web page provides detailed information on the service, statistical data about CODIS records, future plans, and relevant federal legislation about DNA databases in the United States.

Felch, Jason, and Maura Dolan. "How Reliable Is DNA in Identifying Suspects?" *Los Angeles Times*, July 19, 2008. URL: http://articles.latimes.com/2008/jul/20/local/me-dna20. Accessed on March 11, 2009.

The statistical analysis of DNA typing results in a complex and fascinating topic. This article discusses some of the issues involved in making mathematical sense out of such results.

Forensic-Evidence.com. "Master Index." URL: http://www.forensic-evidence.com/site/MasterIndex.html. Accessed on March 11, 2009.

This online newsletter was established to provide a forum for trial lawyers, legal scholars, forensic scientists, students of the law, educators, and public officials to exchange information about the use of scientific techniques in the analysis and litigation of crimes. Some articles available on the site include a discussion of the use of DNA technology in studying remains from the 9/11 disaster, issues in the presentation of DNA evidence to juries, and post-conviction DNA testing.

Human Genome Project. "DNA Forensics." URL: http://www .ornl.gov/sci/techresources/Human_Genome/elsi/forensics .shtml. Accessed on March 11, 2009.

This Web site provides a comprehensive and easily understood introduction to the use of DNA typing in forensic science, including a general introduction to the science and technology involved, some important applications, and some ethical and legal issues involved in the use of DNA typing in forensics.

Zedlewski, Edwin, and Mary B. Murphy. "DNA Analysis for "Minor" Crimes: A Major Benefit for Law Enforcement." *NIJ Journal*, January 2006. URL: http://www.ojp.usdoj.gov/nij/ journals/253/dna_analysis.html. Accessed on March 11, 2009.

DNA typing is most commonly used in cases of murder, assault, or other major crimes. The authors of this article point out that evidence obtained by DNA typing has a broad range of applications and should also be collected in most so-called minor crimes, such as burglary.

Genetic Testing

Books

Betta, Michela, ed. *The Moral, Social, and Commercial Imperatives of Genetic Testing and Screening: the Australian Case.* Dordrecht, The Netherlands: Springer, 2006.

This excellent collection of essays considers the ethical, legal, and commercial aspects of genetic testing, with chapters devoted to topics such as genetic testing in forensic cases, genetic testing and insurance issues, issues involved in community screening programs, the impact of genetic testing on health care for children and parents, genetic testing from a worker's perspective, and the ownership of genetic information. Although the book is couched in terms of experiences in Australia, its analysis and conclusions are relevant to many cultural settings, including that of the United States.

Lemmens, Trudo, Mireille Lacroix, and Roxanne Mykitiuk. *Reading the Future?: Legal and Ethical Challenges of Predictive Genetic Testing.* Montréal: Éditions Thémis, 2007.

The authors examine a number of social, ethical, legal, and other issues raised by the increasing availability of a variety of genetic tests. They consider issues such as the need for genetic testing, regulations that may be necessary or appropriate, stigmatization that may result from certain types of genetic testing results, patenting of genes, commercialization of testing technologies, and clinical issues.

LeVine, Harry. *Genetic Engineering: A Reference Handbook*, 2nd revised edition. Santa Barbara, CA: ABC-CLIO, 2006.

A volume in the publisher's Contemporary World Issues series, this book opens with three chapters devoted to the history and background of genetic engineering, important problems and solutions that have developed in the field, and international perspectives on genetic engineering. The book also includes a number of useful resources for the researchers, including biographical sketches, a chronology of events, organizations involved in the field of genetic engineering, print and electronic resources, and a glossary.

Schoonmaker, Michele, and Erin Dominique Williams. *Genetic Testing: Scientific Background and Nondiscrimination Legislation.* New York: Novinka Books, 2006.

This book is based on a study conducted by the authors for the Congressional Research Service in 2004. It focuses on issues such as the ways in which genetic information differs from other kinds of health information; the ways in which employers use and want to use genetic information; existing and needed government regulations on genetic testing; and implications for a person, his or her family, employers, and others about information obtained by genetic testing.

Sharpe, Neil F., and Ronald F. Carter. *Genetic Testing: Care, Consent, and Liability.* Hoboken, NJ: Wiley-Liss, 2006.

The authors explain the procedure by which genetic testing is conducted and review a number of current applications of the technology. They review a number of related issues, including legal, ethical, psychological, and clinical questions about the use of genetic testing with humans.

Wasserman, David T., Robert Samuel Wachbroit, and Jerome Edmund Bickenbach. *Quality of Life and Human Difference: Genetic Testing, Health Care, and Disability.* Cambridge, UK; New York: Cambridge University Press, 2005.

This collection of nine essays considers a number of ethical issues related to the genetic testing of individuals, especially of newborn children, with special attention to the interpretations of "disability" this technology provides and the issues they raise.

Articles

Berg, Cheryl, and Kelly Fryer-Edwards. "The Ethical Challenges of Direct-to-Consumer Genetic Testing." *Journal of Business Ethics.* 77 (1; 2008): 17–31.

In 2005, 13 Web sites offered genetic testing kits for sale directly to consumers. This articles examines the scientific, legal, and ethical implications of such offers in light of the fact that some kits used questionable and/or unverified scientific procedures and little and/or inadequate information was provided on the use and interpretation of the tests.

Burke, Wylie. "Genetic Testing." *New England Journal of Medicine.* 347 (23; December 5, 2002): 1867–1875.

The author provides a good general introduction to the subject of genetic testing, explaining how the procedure is conducted, the medical benefits it may offer, and some legal and ethical issues involved in its use.

Crockin, Susan L., Gail H. Javitt, Susannah Baruch, and Elizabeth M. Bloom. "Genetic tests are testing the law." *Trial.* 42 (10; October 2006): 44–53.

The technical success of genetic testing has created a number of new legal issues whose resolution has not yet been achieved. This article considers some of those issues and possible ways in which they can be treated.

Hess, Pascale G., H. Mabel Preloran, and C. H. Browner. "Diagnostic Genetic Testing for a Fatal Illness: the Experience of Patients with Movement Disorders." *New Genetics and Society.* 28 (1; March 2009): 3–18.

One of the crucial issues involved in the use of genetic testing is how patients make use of information obtained from the procedure. That issue is especially relevant when the disease being diagnosed is life-threatening. The authors of this article report on a study of 27 neurological consultations and 27 in-depth interviews with patients diagnosed with a fatal disease as a result of genetic testing.

Nicolás, Pilar. " Ethical and Juridical Issues of Genetic Testing: A Review of the International Regulation." *Critical Reviews in Oncology/Hematology.* **69 (2; February 2009): 98–107.**

The author begins with a review of the international legal framework for regulations dealing with genetic testing, primarily the Convention on Human Rights and Biomedicine and the International Declaration on Human Genetic Data, before discussing the objectives of providing protection for patients and the rights of patients involved in genetic testing procedures.

Reports

Council of Europe. *Additional Protocol to the Convention on Human Rights and Biomedicine, Concerning Genetic Testing for Health Purposes.* Strasbourg: Council of Europe, November 27, 2008.

In 1997, the Council of Europe adopted the Convention on Human Rights and Biomedicine, which was later amended four times. The last of these amendments, this document, was adopted after a long discussion of issues related to developments in modern biotechnology as they relate to genetics and human health. So many issues were involved in this discussion that members of the group working on this topic decided to limit their discussions to the topic of genetic testing for health purposes. The final conclusion of those discussions was this document, which is also available online at http://conventions.coe .int/Treaty/EN/Treaties/Html/203.htm.

Holtzman, Neil A., and Michael S. Watson. *Promoting Safe and Effective Genetic Testing in the United States: Final Report of the Task Force on Genetic Testing.* Baltimore: Johns Hopkins University Press, 1998.

The Task Force on Genetic Testing was created in April 1995 by the National Institutes of Health-Department of Energy Working Group on Ethical, Legal, and Social Implications of Human Genome Research. Its charge was to review genetic testing in the United States and to make recommendations that would ensure the development of safe and effective genetic tests. The six chapters in the report deal with a general introduction to the technology and issues raised by its use, ensuring the safety and efficacy of new genetic testing procedures, ensuring the proficiency of testing laboratories, improving the understanding of the technology by health care providers, special problems involved in testing for rare genetic disorders, and summary and recommendations.

Organisation for Economic Co-operation and Development. *Genetic Testing: a Survey of Quality Assurance and Proficiency Standards.* **Paris: Organisation for Economic Co-operation and Development, 2007.**

The report summarizes the results of a survey conducted with 800 genetic testing laboratories in 18 countries in the European Union. It provides information on laboratory policies and practices and the proficiency of workers in those laboratories; policies regarding samples and genetic data handling; and the transborder flow of specimens.

President's Council on Bioethics. *The Changing Moral Focus of Newborn Screening.* **Washington, D.C.: [The President's Council on Bioethics], December 2008.**

Each year, about four million newborn children in the United States are screened for one or more diseases. Technology has made possible screening for a much wider range of diseases, some of which are curable and some are not. This change in technology prompted the President's Council on Bioethics to consider public policy decisions that should be implemented to help parents and health providers to decide which tests under which conditions should be made available to the parents of newborn children.

Internet Sources

Human Genome Project. "Gene Testing." URL: http://www
.ornl.gov/sci/techresources/Human_Genome/medicine/
genetest.shtml. Accessed on March 13, 2009.

This Web page provides a readable general introduction to the
subject of gene testing, including a definition of the procedure,
an explanation as to how it is carried out, some pros and cons
of testing, information on regulation, insurance coverage for the
procedure, and links to other print and electronic resources.

MayoClinic.com. "Genetic Testing." URL: http://www
.mayoclinic.com/health/genetic-testing/MY00370. Accessed on
March 13, 2009.

This highly respected electronic resource provides an overview
of genetic testing for the layperson, focusing on a discussion as
to the conditions under which the procedure is used, how a
patient prepares for a test, what results might be expected, and
how those results are to be assessed and interpreted.

National Human Genome Research Institute. "Frequently
Asked Questions about Genetic Testing." URL: http://www
.genome.gov/19516567. Accessed on March 13, 2009.

This resource provides answers to some of the most commonly
asked and most basic questions about human gene therapy,
including the kind of information available from such tests,
how one decides whether or not to be tested, and the value of
direct-to-consumer commercial genetic test kits. Links to other
print and electronic sources are also available.

National Library of Medicine and National Institutes of
Health. "Genetic Testing." URL: http://www.nlm.nih.gov/
medlineplus/genetictesting.html. Accessed on March 13, 2009.

This is an excellent introductory source on the topic of genetic
testing with basic information on the science and technology
involved, a review of current and recent research, and links to
other print and electronic resources.

Riley, Donald E. "DNA Testing: An Introduction For Non-
Scientists; An Illustrated Explanation." URL: http://www

.scientific.org/tutorials/articles/riley/riley.html. **Accessed on March 13, 2009.**

This Web site provides a very complete and detailed explanation of the technology used in genetic testing. It is written for the general public, although some background in chemistry would be useful for a good understanding of the subject matter. Probably an essential resource for anyone who wishes to understand the technical basis of genetic testing.

Human Gene Therapy

Books

Crommelin, D. J. A., Robert D. Sindelar, and Bernd Meibohm. *Pharmaceutical Biotechnology: Fundamentals and Applications.* **New York: Informa Healthcare, Inc., 2008.**

Chapter 1 of this book provides a good general introduction to recombinant DNA techniques at a somewhat technical level, while chapter 7 deals specifically with the methods and applications of human gene therapy.

Dropulic, Boro, and Barrie Carter, eds. *Concepts in Genetic Medicine.* **Hoboken, NJ: Wiley-Liss, 2008.**

This collection of articles deals with advances in the technology of gene therapy as well as a review of progress in its use in treating a number of diseases. Regulatory, legal, and economic issues are also considered.

Huber, Brian E., and Ian Magrath, eds. *Gene Therapy in the Treatment of Cancer: Progress and Prospects.* **Cambridge: Cambridge University Press, 2007.**

This book is a collection of 10 papers dealing with recent developments in the use of gene therapy for the treatment of cancers. The focus is primarily on vectors used in gene therapy, their efficacy and safety.

Kelley, Evelyn B. *Gene Therapy.* **Westport, CT: Greenwood Press, 2007.**

This book is divided into three sections, the first of which provides a broad, general introduction to the principles and methods of gene therapy. Section 2 focuses on issues relating to the use of gene therapy to treat diseases in humans. Section 3 is a collection of ancillary materials useful to anyone pursuing further research in the subject, such as a chronology of important events, biographical sketches of important figures in the field, a list of bibliographical resources, a glossary, and a list of organizations interested in gene therapy.

Meyers, Robert A. *Pharmacology : from Drug Development to Gene Therapy.* Weinheim, Germany: Wiley-VCH, 2008.

Part IV of this two-volume work is a selection of articles dealing with the status and methods of modern gene therapy techniques.

Panno, Joseph. *Gene Therapy: Treating Disease by Repairing Genes.* New York: Facts On File, 2008.

This book is part of Fact On File's New Biology series, designed for high school students and general readers. It provides a review of the methodology and applications of gene therapy along with a number of research aids, including a chronology, set of biographical sketches, list of print and electronic resources, glossary, and list of organizations involved in gene therapy.

Articles

Aiuti, A., et al. "Progress and Prospects: Gene Therapy Clinical Trials (Part 2)." *Gene Therapy.* 14 (22; November 2007): 1555–1563.

Alexander, B. L., et al. "Progress and Prospects: Gene Therapy Clinical Trials (Part 1)." *Gene Therapy* 14 (20; October 2007): 1439–1447.

These two articles contain commentaries by 20 leading researchers in the field of human gene therapy about progress in the technology over the preceding two decades in treating a variety of diseases, including cystic fibrosis, Duchenne muscular dystrophy, cancer, cardiovascular disorders, and severe combined immunodeficiency disease.

An important resource for reports on recent developments in the science and technology of human gene therapy is the monthly journal of that name, *Human Gene Therapy*, published by Mary Ann Liebert, Inc., ISSN 1043–0342; online ISSN: 1557–7422.

Castanotto, Daniela, and John J. Rossi. "The Promises and Pitfalls of RNA-interference-based Therapeutics." *Nature.* 457 (January 22, 2009): 426–433.

The authors discuss a relatively new approach to the control of gene expression in cells, a procedure which provides another method of conducting gene therapy. They describe progress in the field along with possible future applications.

Herrmann, F. "Clinical Application of Gene Transfer." *Journal of Molecular Medicine.* 74 (4; April 1996): 213–221.

This article is of some historical interest because it was written in the early days of gene transfer experiments, when some early clinical trials were underway. The perspective presented in the article should be compared with the eventual results of these experiments and later attempts to complete clinical trials of gene therapy experiments.

Kimmelman, Jonathan. "The Ethics of Human Gene Transfer." *Nature Reviews Genetics.* 9 (3; March 2008): 239–244.

The author reviews briefly the scientific and technological history of gene therapy and then points out that some fundamental ethical problems have still not been resolved about the use of the technology. Such questions include the decision as to when clinical trials with humans should be started, the acceptability of modifying a person's genome, and the issue as to whether the technology should be used for the enhancement of traits in addition to the cure of diseases.

Kittredge, Clare. "Gene Therapy. Under the Microscope Again." *The Scientist.* 19 (9; 2005): 14–17.

The author reviews some of the failed experiments that have plagued gene therapy experiments over the past two decades, but points out the success that has also been achieved. She considers the future of the technology as a method of treating diseases.

Preuss, Meredith A., and David T. Curiel. "Gene Therapy: Science Fiction or Reality?" *Southern Medical Journal.* **100 (1; January 2007); 101–104.**

The authors review the scientific principles involved in human gene therapy, offer a brief review of its historical development, discuss recent developments in the technology of gene therapy, and lay out some possible future applications of the technology.

Russell, Stephen J., and Kah-Whye Peng. "Primer on Medical Genomics. Part X: Gene Therapy." *Mayo Clinic Proceedings.* **78 (11; November 2003): 1370–1383.**

The authors provide a comprehensive review of the nature of gene therapy and its past successes and failures. It outlines the potential applications of the technology and some safety issues related to its application.

Reports

Edgar, Harold, and Thomas Tursz. *Report on Human Gene Therapy.* **[Paris]: United Nations Educational, Scientific, and Cultural Organization, December 1994.**

One of the earliest reports on gene therapy, this document was prepared as the result of a meeting of UNESCO's International Bioethics Committee in September 1994. It reviews the technologies used in gene therapy and a number of ethical, legal, and other issues that are raised by availability of the technology.

Gene Therapy Advisory Committee. *Fourteenth Annual Report: Covering the Period from January 2007 to December 2007.* **London: Health Department of the United Kingdom, 2008.**

Every year since 1995, the British Health Department has been publishing a report of applications made for gene therapy research and clinical trials. As of the 14th report in the series, the department has considered 155 applications, of which it has approved 126. Each year's report summarizes the applications that have been made, the actions that have been taken, and the results of clinical trials and other research. This series of reports provides an invaluable resource on the research being conducted in the United Kingdom on human gene therapy.

Jain, K. K. *Special Report on Gene Therapy Companies*, **2 vols. Chichester, England; New York: John Wiley & Sons, 2009.**

This report occurs regularly (usually annually) and begins with an extended description of current technologies used in gene therapy as well as a review of recent clinical trials in the field. It then profiles companies (192 in the 2009 edition) involved in the development of gene therapy materials and methods. The book concludes with 670 references to sources on gene therapy.

National Reference Center for Bioethics Literature. Joseph and Rose Kennedy Institute of Ethics. Georgetown University. *Human Gene Therapy.* **Washington, D.C.: Georgetown University, August 2008.**

This report was first published in March 1994 and has been updated regularly ever since, most recently in August 2008. It provides a comprehensive overview of the field of human gene therapy in a format easily comprehensible to the average reader. It includes a general introduction to the topic, a history of gene therapy in the United States, a discussion of techniques, arguments in favor of and opposed to gene therapy, and a list of organizations and references on the topic of gene therapy.

Office of Technology Assessment. *Human Gene Therapy—A Background Paper.* **Washington, D.C.: Office of Technology Assessment, December 1984.**

Although this report is now more than 20 years out of date, it remains a remarkably useful general introduction to the subject of human gene therapy, the technologies available (at the time), the medical opportunities available with the use of gene therapy, reasons for governmental concern about gene therapy, and the social and ethical issues posed by use of the technology. The report is available online at http://govinfo.library.unt.edu/ota/Ota_4/DATA/1984/8415.PDF.

Orkin, Stuart H., and Arno G. Motulsky. *Report and Recommendations of the Panel to Assess the NIH Investment in Research on Gene Therapy.* **n.p.: December 7, 1995. Available**

online. URL: http://www.nih.gov/news/panelrep.html. Accessed on March 11, 2009.

This report was prepared at the request of Dr. Harold Varmus, then-director of the National Institutes of Health, for an assessment of the federal government's role in research on human gene therapy. The report summarizes the conclusions reached as the result of three meetings held with experts in the area in May, July, and August of 1995.

Internet Sources

Gene Therapy Net.com. URL: http://www.genetherapynet.com/index.html. Accessed on March 11, 2009.

This Web site is a resource for professionals and patients who are interested in the subject of genetic therapy. It provides a general introduction to the procedures by which gene therapy is conducted, recent and current research in the field, organizations and industries involved in the process, regulatory issues, clinical trial databases, and suggestions for further reading and research.

Human Gene Project. "Gene Therapy." URL: http://www.ornl .gov/sci/techresources/Human_Genome/medicine/genetherapy .shtml. Accessed on March 11, 2009.

This Web page provides a general, well-written, easily understood introduction to the subject of human gene therapy. It explains how the procedure is carried out, the diseases for which it might be used, recent developments in the field, and controversial issues related to the technology.

Kids Health for Parents. "Gene Therapy and Children." URL: http://kidshealth.org/parent/system/medical/gene_therapy .html. Accessed on March 17, 2009.

This article reviews some basic information about human gene therapy and then discusses some special issues involved in the use of the technology among children.

Medline Plus. "Genes and Gene Therapy." URL: http:// www.nlm.nih.gov/medlineplus/genesandgenetherapy.html. Accessed on March 11, 2009.

A project of the U.S. Library of Medicine and the National Institutes of Health, this Web page provides a comprehensive review of genetic therapy that includes the basic science and technology involved, a review of current research, and links to other resources on the topic.

PubMed. "Gene Therapy." URL: http://www.ncbi.nlm.nih.gov/ sites/entrez?cmd=search&db=pubmed&term=gene+therapy [majr]+AND+humans[mh]+AND+english[la]+AND+(review [pt]+OR+guideline[pt]+OR+clinical+trial[pt]+OR+jsubsetk [text]+OR+jsubsetaim[text]+OR+jsubsetn[text]+OR+patient +education+handout[pt]+NOT+(letter[pt]+OR+editorial[pt] +OR+case+reports[pt]))&doptcmdl=summary&cmd_current= Limits&pmfilter_EDatLimit=last+1+Year&tool=MedlinePlus. Accessed on March 11, 2009.

PubMed is a service of the U.S. Library of Medicine and the National Institutes of Health that provides a superb guide to articles on virtually any topic in health and medicine. The site is also connected to PubMed Central, a source of full-text articles available at no cost on the Web. The above Web address leads to a select group of articles of special interest to readers of this book, but a less restrictive search ("Gene Therapy") produces a much larger list of articles. Readers may be interested in gene therapy research on specific diseases, many of which are included in the above search. These articles tend to be very technical, but may provide a good general introduction to progress in specific fields, such as diabetes, cancer, or heart disease.

Genetically Modified Foods

Books

Byrne, Jay, et al. *Let Them Eat Precaution: How Politics Is Undermining the Genetic Revolution in Agriculture.* **Washington, D.C.: AEI Press, 2006.**

The authors argue that modern science has produced a critical breakthrough in solving the world's food shortage by the development of genetically modified foods that are nutritious, environmentally safe, and relatively inexpensive. They are that the full potential of this breakthrough is not being realized, however, because of campaigns by advocacy groups who worry about potential risks of GM products and, probably more important, about the potential takeover of the world's methods of food production by large corporate entities. They suggest some ways in which accommodations can be reached between proponents and critics of GM foods to achieve a better utilization of this new technology.

Evenson, R. E., and V. Santaniello, eds. *Consumer Acceptance of Genetically Modified Foods.* **Wallingford, UK; Cambridge, MA: CABI Publications, 2004.**

Scientists, politicians, environmentalists, consumer advocates, and others argue about the safety of GM foods and their role in today's world. But the bottom line is whether or not consumers will actually buy and use such foods. This book includes a number of papers presented at the International Consortium on Agricultural Biotechnology Research, held in Ravello, Italy, in July 2002 dealing with this question.

Fedoroff, Nina V., and Nancy Marie Brown. *Mendel in the Kitchen: A Scientist's View of Genetically Modified Foods.* **Washington, D.C.: Joseph Henry Press, 2004.**

A geneticist and researcher in biotechnology (Fedoroff) and a science writer (Brown) combine to review the development of genetically modified foods, their place in the human diet today, claims of risks posed by GM foods, and the validity of those claims.

Henningfeld, Diane Andrews. *Genetically Modified Food.* **Detroit: Greenhaven Press, 2009.**

This book is written for young adults. It provides a general introduction to the science and technology of GM foods and the social, economic, ethical, legal, and political issues raised by their production and consumption.

Smith, Jeffrey M. *Genetic Roulette: The Documented Health Risks of Genetically Engineered Foods.* **Fairfield, IA: Yes! Books, 2007.**

The author lists nearly 40 health risks he claims are associated with genetically modified foods, provides some general information about each risk, and documents the basis for his statement with detailed references from the literature about the topic. The second half of the book is devoted to his argument that children are most at risk from genetically modified foods and his reasons for believing that contention.

Thomson, Jennifer A. *Seeds for the Future: The Impact of Genetically Modified Crops on the Environment.* Ithaca, NY: Comstock Publishing Associates, 2007.

The author argues that the wider use of GM foods may be the only long-term answer to feeding the world's growing population. She acknowledges the ecological risks posed by the cultivation of such crops, analyzes the basis for these concerns, and outlines the likelihood that GM crops may result in widespread environmental disasters in areas where they are grown or consumed.

Weasel, Lisa H. *Food Fray: Inside the Controversy over Genetically Modified Food.* New York: Amacom-American Management Association, 2009.

The author provides a general introduction to the current controversy over the production of genetically modified foods. The scientific background is presented at a level that most readers will be able to understand. The focus of the book is a review of a number of important instances involving the production and use of GM foods, such as the use of such foods to deal with food shortages in Africa, the genetic engineering of the milk supply in the United States, the development and uses of GM papaya, and the battle over the use of GM foods in India.

Articles

Conko, Greg. "Modified Crops—Regulate the Product, Not the Process." *Chemical Engineering Progress.* 101 (9; September 2005): 4–5.

The author argues that the use of modern biotechnological techniques to modify crops is actually safer than traditional "hit-or-miss" techniques. The article is half of a Critical Issues Forum discussion in this issue of the magazine. Also see Shave and Azeez in this section below.

Dona, Artemis, and Ioannis Arvanitoyannis. "Health Risks of Genetically Modified Foods." *Critical Reviews in Food Science and Nutrition.* 49 (2; February 2009): 164–175.

The authors argue that health effects of GM foods on humans are numerous and well-documented. They suggest that much more research is needed on this issue, but that such research should focus on the effects of GM foods on body organs rather than on the risks posed by specific GM foods, as has often been the case in the past.

Knight, John G., David K. Holdsworth, and Damien W. Mather. "GM Food and Neophobia: Connecting with the Gatekeepers of Consumer Choice."*Journal of the Science of Food and Agriculture.* 88 (5; April 15, 2008): 739–744.

The acceptance of genetically modified foods in some countries has remained low for at least two reasons: government regulations limiting the production, importation, and sale of such foods; and neophobia, the irrational fear of risks posed by such foods. The authors claim that GM foods hold the potential for solving food problems in many nations and that, eventually, pragmatic forces will overcome neophobia and increase the likelihood of many nations accepting a greater role for GM foods in their people's diets.

Shave, John, and Gundula Azeea. "GM Crops—Not All Sugar and Spice." *Chemical Engineering Progress.* 101 (September 9, 2005): 6–7.

The authors argue that the production of genetically modified foods is producing a host of biological, environmental, economic, and legal problems. The article is half of a Critical Issues Forum discussion on GM foods. Also see Conko in this section above.

Strauss, Debra M. "Feast or Famine: The Impact of the WTO Decision Favoring the U.S. Biotechnology Industry in the EU Ban of Genetically Modified Foods." *American Business Law Journal.* 45 (4; 2008): 775–826.

One of the most important political disputes over GM foods in recent years has been the effort of the United States, Canada, and other nations to sell such foods to consumers in the European Union, which imposed a five-year moratorium on such foods in 1999. In 2003, these nations presented a formal complaint to the World Trade Organization about this ban. This article considers the issues involved in this complex and very important debate about GM foods.

Reports

Committee on Environmental Impacts Associated with Commercialization of Transgenic Plants. Board on Agriculture and Natural Resources. National Research Council. *Environmental Effects of Transgenic Plants: The Scope and Adequacy of Regulation.* **Washington, D.C.: National Academies Press, 2002.**

This book reports on a survey of current practices in the development and use of transgenic plants, with possible risks they may pose to the environment and to human health. The committee reviews current federal legislation dealing with transgenic plants and recommends additional safeguards.

Committee on Genetically Modified Pest-Protected Plants. National Research Council. *Genetically Modified Pest-Protected Plants: Science and Regulation.* **Washington, D.C.: National Academies Press, 2000.**

A major area of recombinant DNA research is the development of new plant forms that are resistant to pests. Some critics have argued that such plants may pose a threat both to human health and to the natural environment. This report summarizes the information available on this subject and recommends regulation that may be helpful in ensuring the safety of pest-protected plants in the future.

Committee on Identifying and Assessing Unintended Effects of Genetically Engineered Foods on Human Health. National Research Council. *Safety of Genetically Engineered Foods: Approaches to Assessing Unintended Health Effects.* **Washington, D.C.: National Academies Press, 2004.**

A group of experts representing the Food and Nutrition Board, Institute of Medicine, Board on Agriculture and Natural Resources, Board on Life Sciences, and Earth and Life Studies studied the potential health risks posed by the development of genetically engineered foods. They concluded that those foods posed no more of a threat than do foods processed by any other commercially-available procedure. Any possible health issues must be assessed on a case-by-case basis, the committee said.

Friends of the Earth, International. *Who Benefits From GM Crops? Feeding the Biotech Giants, Not the World's Poor.* **Amsterdam: Friend of the Earth, February 2009.**

This report presents a very detailed, well-documented case that the greatest beneficiaries of the production of GM foods is not consumers, but food production companies. The report presents statistical data on a host of GM-related issues, including GM products now in production and use; percentage of land devoted to GM crop production; amount and cost of pesticides used in GM crop production; average cost of various GM foods in the United States and other nations; and crop yields for GM and conventional foods.

Pool, Robert, and Joan Esnayra. *Ecological Monitoring of Genetically Modified Crops: A Workshop Summary.* **Washington, D.C.: National Academies Press, 2001.**

This book is a summary of a conference held on July 13–14, 2000, on the ecological effects of genetically modified crops. The workshop did not deal with human health effects of GM foods. The purpose of the workshop was to identify issues involved in monitoring GM crops, describing what is already known about the subject, and identifying needed areas of future research.

U.S. General Accounting Office. *Genetically Modified Foods. Experts View Regimen of Safety Tests as Adequate, but FDA's Evaluation Process Could Be Enhanced.* **Washington, D.C.: U.S. General Accounting Office, May 2002.**

At the request of U.S. Representatives John E. Baldacci and John F. Tierney, the U.S. General Accounting Office (now the U.S. General Accountability Office; GAO), conducted a study on a number of issues associated with the production and consumption of GM foods, including health risks that may be associated with such foods, regulatory mechanisms currently employed by the Food and Drug Administration (FDA) with regard to such foods, the future of GM foods and tests that may be needed to ensure their safety, and expert views on the feasibility of determining long-term health problems that may be associated with contact with GM foods. The GAO's general conclusion is reflected in the title of the report.

Vogt, Deborah U., and Mickey Parish. *Food Biotechnology in the United States: Science, Regulation, and Issues.* **Washington, D.C.: Congressional Research Service, January 19, 2001.**

This report reviews the risks and benefits of genetically-modified foods, with special attention to issues such as the science of genetic engineering, federal regulations dealing with GM foods, and current issues relating to the production and sale of GM foods.

Internet Sources

FoodRisk.org. "Genetically Modified Foods." URL: http:// www.foodrisk.org/commodity/genetically_modified/index.cfm . Accessed on March 14, 2009.

A service of the Joint Institute for Food Safety and Applied Nutrition at the University of Maryland, this Web site provides research articles, reports, and other documents dealing with food-related risks, including the use of genetically modified foods. The Web site is divided into eight sections dealing with hazard identification, risk and safety assessment, risk management, risk communication, labeling, GM foods and animal products. GM foods and plant products, and other GM foods resources.

International Service for the Acquisition of Agri-biotech Applications. "Biotech Crops Experience Remarkable Dozen Years of Double-Digit Growth: Socio-economic Benefits Becoming Evident among Resource-poor Farmers." URL: http://www.isaaa.org/resources/Publications/briefs/37/press-release/default.html. Accessed on March 14, 2009.

This press release summarizes some of the data about increases in GM food production and use around the world for the year 2007. The article claims that business is doing very well, at least in part because poor farmers are beginning to understand and appreciate the benefits of farming with GM crops and related products.

Monsanto. "Conversations about Plant Biotechnology." URL: http://www.monsanto.com/biotech-gmo/asp/default.asp. Accessed on March 14, 2009.

One of the world's largest producers of genetically modified foods and related products maintains this Web site, which contains a number of video presentations with farmers who use Monsanto technology and scientists who are experts in the field of GM foods, as well as similar presentations on a number of topics related to genetically modified foods. The Web site also has a page containing pertinent facts on the production and use of GM foods.

Pusztai, Arpad. "Genetically Modified Foods: Are They a Risk to Human/Animal Health?" ActionBioscience.org. URL: http://www.actionbioscience.org/biotech/pusztai.html. Accessed on March 14, 2009.

This is a very good review of the scientific information available (at the time the article was published) about safety hazards related to the production and consumption of genetically modified foods. The author agrees with an article in the journal *Science* claiming that there have been "many opinions but few data" on the question. The article concludes with a number of excellent references.

Smith, Jeffrey M. "Point of View: Genetically Modified Foods Unsafe? Evidence that Links GM Foods to Allergic Responses Mounts." Genetic Engineering and Biotechnology News. URL: http://www.genengnews.com/articles/chitem.aspx?aid=2252. Accessed on March 14, 2009.

The author suggests that studies have shown that GM foods are associated with a number of medical problems, including allergic responses, damage to the immune system, and toxicity and reproductive problems. His bibliography lists 43 studies to support his views. The Smith article generated a considerable response on the Web site's blog, which can be accessed at http://www.genengnews.com/blog/item.aspx?id=93.

Vithani, Neha, and Jigar N. Shah. 'Transgenic Animals—A Boon By Biotechnology." URL: http://www.pharmainfo.net/

reviews/transgenic-animals-boon-biotechnology. Accessed on April 7, 2009.

The authors present a comprehensive, detailed, moderately technical, and very well written overview of transgenic animals, providing a nice historical background, a relatively simple explanation of the technology involved, many potential applications of transgenic animals, and some risks and benefits associated with their use.

Weissman, Robert. "Heads Monsanto Wins, Tails We Lose; The Genetically Modified Food Gamble." URL: http://www .commondreams.org/archive/2008/03/19/7769. Accessed on March 14, 2009.

The author reviews the recent explosion in genetically modified food production and suggests some cautions to be observed in its use in human diets. The article is written in response to the report by the International Service for the Acquisition of Agribiotech Applications listed above.

Whitman, Deborah B. "Genetically Modified Foods: Harmful or Helpful?" ProQuest. URL: http://www.csa.com/discovery guides/gmfood/overview.php. Accessed on March 14, 2009.

This article presents an excellent, well-balanced overview of genetically modified foods, with separate sections on the advantages of GM foods, their current prevalence in agricultural production, some criticisms about the production and consumption of GM foods, governmental regulations of the production and sales of such foods, and the labeling of GM foods and foods that contain genetically modified materials. The author provides an extensive and very useful bibliography on the topic.

World Health Organization. "20 Questions on Genetically Modified (GM) Foods." URL: http://www.who.int/foodsafety/ publications/biotech/en/20questions_en.pdf. Accessed on March 14, 2009.

The world's leading health authority provides answers to a number of fundamental questions about genetically modified foods, including what the term means, which foods have been genetically modified thus far, what risks and benefits can be attributed to GM foods, how GM foods are regulated, and the debate over GM foods in the European Union and other parts of the world.

Pharming

Books

Engelhard, Margaret, Kristin Hagen, and Felix Thiele, eds. *Pharming: A New Branch of Biotechnology.* Bad Neuenahr-Ahrweiler: Europäische Akademie, 2007.

This work is publication number 43 in the academy's Graue Reihe (Gray Series) of studies on the individual, social, and environmental consequences of developments in science and technology. The five papers in the book deal with a general introduction to the history and technology of pharming, the use of plants and animals for the production of pharmaceuticals, and ethical and legal issues raised by the development and use of pharming techniques.

Englehard, Margaret, Kristin Hagen, and Mathias Boysen, eds. *Genetic Engineering in Livestock: New Applications and Interdisciplinary Perspectives.* Berlin: Springer, 2009.

This collection of six articles provides a general overview of the subject of pharming, with special attention to the techniques by which pharming is conducted, past experience and future prospects for the technology, ethical issues, and issues concerning the welfare of pharmed animals.

Faye, Loïc, Véronique Gomord, and Jean-Philippe Salier. *Recombinant Proteins from Plants: Methods and Protocols.* New York: Humana, 2009.

A publication in the Methods in Molecular Biology series, this book is a technical work describing the detailed procedures by which pharming experiments are conducted.

Rasco, Eufemio Tam. *The Unfolding Gene Revolution: Ideology, Science, and Regulation of Plant Biotechnology.* [Manila] : International Service for the Acquisition of Agri-Biotech

Applications: Southeast Asian Regional Center for Graduate Study and Research in Agriculture, 2008.

The author reviews recent advances in the biotechnology of plants, with special attention to the development of pharming technologies, and considers issues related to the regulation of research and application of such technologies.

Rehbinder, Eckard, et al. *Pharming: Promises and Risks of Biopharmaceuticals Derived from Genetically Modified Plants and Animals.* Berlin: Springer, 2009.

This text provides a general introduction to the subject of pharming, beginning with an explanation of the technology and followed by a discussion of issues such as risk assessment of plant and animal pharming, the welfare of pharming animals, public attitudes about pharming, legal issues related to pharming practices, and patent issues related to the development of pharming technologies.

Articles

Goldstein, D. A., and J. A. Thomas. "Biopharmaceuticals Derived from Genetically Modified Plants." *QJM*. 97 (11; November 2004):705–16.

This review article discusses the development of pharming technologies and their possible applications in the future. The authors point out that product safety issues for patients, workers, and the general public must be resolved and appropriate regulatory standards must be adopted before commercialization of pharming technologies can be achieved.

Kaiser, Jocelyn. "Is the Drought Over for Pharming?" *Science*. 320 (5875; April 25, 2008): 473–475.

The author cites data that suggests that biotechnology companies are moving ahead aggressively in the development of technologies to manufacture drugs and other compounds using plants. Although technological, ethical, social, and legal questions remained to be answered, these issues do not seem to have interrupted the flow of new products from the industry.

Kind, Alexander, and Angelika Schnieke. "Animal Pharming, Two Decades On."*Journal Transgenic Research.* 17 (6; December 2008): 1025–1033.

The authors point out that scientists have been touting the advantages of pharming for at least two decades, but the first product from the technology was not approved for use until 2006. They discuss some of the problems involved in making pharming a realistic practical alternative to existing methods of drug production and the kinds of products that may be most suitable for this method of production.

Liénard, David, Christophe Sourrouille, Véronique Gomord, and Loïc Faye. "Pharming and Transgenic Plants." *Biotechnology Annual Review.* 13 (2007): 115–147.

This article reviews the historic role of plants in the production of pharmaceuticals and points out that recombinant DNA procedures have restored the significance of plants in this process. The authors review the techniques that have been developed for the manufacture of proteins and other essential compounds by plants, as well as the specific types of compound production that are now possible with the technology.

Twyman, Richard M., Stefan Schilberg, and Rainer Fisher. "Transgenic Plants in the Biopharmaceutical Market." *Expert Opinion on Emerging Drugs.* 10 (1; February 2005): 185–218.

The authors point out that the potential of genetically modified plants as the source of drugs has increased dramatically in recent years. In this article, they describe some of the technologies that have been developed for the pharming of plants and compares the efficacy and safety of these technologies to more traditional methods of drug manufacture. They also discuss a number of nonscientific issues that must be resolved before the pharming of drugs becomes a practical reality.

Reports

Pew Initiative on Food and Biotechnology. *Pharming the Field: A Look at the Benefits and Risks of Bioengineering Plants to Produce Pharmaceuticals.* Washington, D.C.: The Pew Charitable Trusts, 2003.

This document reports on the results of a workshop sponsored by the Pew Initiative on Food and Biotechnology, the U.S. Food and Drug Administration, and the Cooperative State Research, Education and Extension Service of the U.S. Department of Agriculture held in Washington in July 2002. The report summarizes potential risks and benefits associated with the technology, along with regulatory and legal issues that arise as a result of its use in the production of pharmaceuticals.

U.S. Department of Agriculture. Animal and Plant Health Inspection Service. Biotechnology Regulatory Services. *Guidance for APHIS Permits for Field Testing or Movement of Organisms Intended for Pharmaceutical or Industrial Use.* **URL: http:// www.aphis.usda.gov/brs/pdf/Pharma_Guidance.pdf. Accessed on March 15, 2009.**

The USDA Biotechnology Regulatory Services (BRS) agency is responsible for monitoring all experiments that involve the introduction of foreign genes into plants for the purpose of making compounds of industrial, pharmaceutical, or other applications. BRS has prepared this document to explain to researchers the steps required in applying for and receiving the license needed to carry out such experiments.

U.S. Food and Drug Administration. *Guidance for Industry: Drugs, Biologics, and Medical Devices Derived from Bioengineered Plants for Use in Humans and Animals.* **Washington, D.C.: U.S. Food and Drug Administration, September 2002.**

This document was prepared jointly by the U.S. Department of Agriculture and the Food and Drug Administration to alert producers of "drugs, biologics, and medical devices derived from bioengineered plants" as to the information they will be expecting to receive in anticipation of the granting of licenses for research in this field.

Wach, Michael J. Biotechnology Regulatory Services. *Introduction of Genetically Engineered Organisms: Draft Programmatic Environmental Impact Statement.* **Washington, D.C.: U. S Department of Agriculture. Marketing and Regulatory Programs. Animal and Plant Health Inspection Service, July 2007.**

The USDA Animal and Plant Health Inspection Service is the federal agency primarily responsible for writing and enforcing regulations dealing with the introduction of genes into foreign organisms. Its authority to regulate bioengineered plants resides in the Plant Protection Act of 2000. This document announces the agency's intention to consider the revision of regulations dealing with the genetic manipulation of plants, including the production of new plant forms for the purpose of pharming. The general outlines of the agency's intentions are provided in this document, although final regulations had not yet been published as of mid-2009.

Internet Sources

Department of Soil and Crop Sciences. Colorado State University. "Transgenic Crops: An Introduction and Resource Guide." URL: http://www.cls.casa.colostate.edu/TransgenicCrops/index.html. Accessed on March 15, 2009.

A superb resource on the subject of pharming that includes a history of plant breeding, explanations of the technology involved in making transgenic plants, a discussion of regulation, and current and future transgenic crops. The Web site has not been updated since 2004, but still provides some very valuable basic information on the subject.

Gillespie, David. "Pharming for Pharmaceuticals." URL: http://learn.genetics.utah.edu/archive/pharming/index.html. Accessed on March 15, 2009.

The author provides a clear, understandable introduction to the technology of pharming and reviews some of the ethical concerns posed by its applications.

Ho, Mae-Wan, and Joe Cummins. "Molecular Pharming by Chloroplast Transformation." URL: http://www.i-sis.org.uk/MPBCT.php. Accessed on March 15, 2009.

The authors describe a procedure by which the chloroplast in plants can be transformed to create an organism that produces specific compounds, such as those used in making drugs and medicines. They point out some of the environmental problems inherent in this procedure, which has become quite popular among industrial pharmaceutical companies.

International Academy of Life Sciences. "Introduction to Plant-Made Pharmaceuticals." URL: http://www.plantpharma .org/plant-made-pharmaceuticals/introduction-to-plant-made-pharmaceuticals/. Accessed on March 15, 2009.

PlantPharma is an online community of scientists from around the world interested in the development of technologies for the production of pharmaceuticals by the genetic engineering of plants. This Web site provides information about the organization and its work, including sections on latest news in the field, description of ongoing research, reports on patents, and explanations of procedures used in pharming experiments.

Katsnelson, Alla. "Fear of Pharming: Controversy Swirls at the Crossroads of Agriculture and Medicine." Scientific American. URL: http://www.sciam.com/article.cfm?id=fear-of-pharming. Accessed on March 15, 2009.

The author reviews the history and growth of pharming technology and points out reasons that some critics are concerned about possible effects on human health and environmental quality. He then discusses the process by which pharmed plants are regulated in the United States and some problems that may remain in the control of this technology.

Pharma-Planta. URL: http://www.pharma-planta.org/index.htm. Accessed on March 15, 2009.

Pharma-Planta is a consortium of 39 scientists from academic institutions in Europe and South Africa, funded by the European Commission to develop efficient and safe technologies for the use of plants to develop pharmaceuticals. This Web page contains detailed information on the group's current research efforts, technologies it is investigating and/or using, links to related resources, and a collection of news articles about its efforts.

Cloning

Books

Cibelli, Jose B., Robert P. Lanza, Keith Campbell, and Michael D. West, eds. *Principles of Cloning*. Amsterdam; Boston: Academic Press, 2008.

This volume is a technical presentation of the fundamental principles of cloning that may be understandable, at least in part, to a general audience. An effort to comprehend most of the papers in the book is well worth the reader's while. Topics in the book are organized under a few general rubrics, including a historical perspective, basic biological procedures, cloning by species (amphibians, fish, mice, rabbits, swine, cattle, etc.), nuclear transfer in primates, applications of cloning, and legal and ethical issues raised by cloning experiments and applications.

Klotzko, Arlene Judith. *A Clone of Your Own?: The Science and Ethics of Cloning.* **New York: Cambridge University Press, 2006.**

After a general introduction to the science and technology of cloning, the author focuses on three areas of particular ethical concern in cloning studies: the cloning of farm animals, therapeutic cloning, and reproductive cloning.

Levine, Aaron D. *Cloning: A Beginner's Guide.* **Oxford, UK: Oneworld, 2007.**

This is a clearly written introduction to the basics of cloning with chapters that cover the scientific principles of the technology, a history of cloning experiments in the twentieth century, ethical issues related to the practice of cloning, and future prospects of the technology.

Roleff, Tamara L. *Cloning.* **San Diego: ReferencePoint Press, 2009.**

This book is part of Greenhaven Press' long-standing Opposing Viewpoint series in which two or more positions on important issues are provided. The book is intended for readers in grade 9 and up.

Wilmut, Ian, and Roger Highfield. *After Dolly: The Uses and Misuses of Human Cloning.* **New York: W. W. Norton, 2006.**

One of the authors of this book (Wilmut) was lead scientist in the project to clone the first mammal, a sheep named Dolly, in 1996. This book reviews that accomplishment and discusses a number of ethical issues raised by it, including the cloning of animals other than humans, cloning for therapeutic reasons, "designer babies," and the reasons scientists should make no efforts to clone human beings.

Articles

Birmingham, Karen. "The Move to Preserve Therapeutic Cloning." *Journal of Clinical Investigation.* **112 (11; December 1, 2003): 1600–1601.**

The author takes note of recent claims by some scientists that they are studying the possibilities of human reproductive cloning and the political consequences of such claims. She reviews United Nation's actions (and inactions) on this issue to comment on the dilemma faced by scientists in rejecting the possibility of research on reproductive cloning while maintaining the right to work on therapeutic cloning.

Heatherington, Tracey. "Cloning the wild mouflon." *Anthropology Today.* **24 (1; January 2008): 9–14.**

The author describes a successful experiment in which the critically endangered mouflon is cloned. She then considers some issues for the future of conservation biology posed by the success of this procedure.

Kfoury, Charlotte. "Therapeutic Cloning: Promises and Issues." *McGill Journal of Medicine.* **10 (2; July 2007): 112–120.**

The author points out the very wide range of potential benefits accruing from the development of successful therapeutic cloning procedures, but then reviews some of the problems associated with such work, including legal and ethical issues that need to be resolved and separating therapeutic from reproductive cloning.

Miller, H. I. "Fear and Cloning." *Regulation.* **31 (2; June 22, 2008): 5–9.**

The author comments on the decision by the U.S. Food and Drug Administration that food products from cloned animals pose no health threat to consumers. He complains, however, that the decision was so long in coming and that a contrary opinion was issued almost simultaneously by another government agency, the U.S. Department of Agriculture.

Strong, C. "Reproductive Cloning Combined with Genetic Modification." *Journal of Medical Ethics.* 31 (11; November 2005): 654–658.

The author reviews the objections to human reproductive cloning and points out how modifications in the standard procedure could be made (genetic modification) to perhaps make the technology more acceptable. He suggests that the modified procedure might be of value to infertile, gay, and/or lesbian couples who would normally not procreate by traditional methods available to fertile heterosexual couples.

Reports

Committee on Science, Engineering, and Public Policy, et al. *Scientific and Medical Aspects of Human Reproductive Cloning.* Washington, D.C.: National Academy Press, 2002.

This report consists of six chapters which discuss the science and technology of cloning and its relation to stem cell research, an overview of the status of research on animal cloning and its potential applications to human cloning, a review of current technologies for assisted reproduction, a discussion of plans by those who hope to pursue research on human cloning, and the panel's recommendations for public policy guidelines on human cloning.

Cowan, Tadlock, and Geoffrey S. Becker. *Biotechnology in Animal Agriculture: Status and Current Issues.* Washington, D.C.: Congressional Research Service, February 11, 2009.

This report reviews the current status of recombinant DNA research on animals, regulation and oversight, international trade issues, and policy concerns. The report focuses on five technologies in particular: embryo transfer, in vitro fertilization, sexing embryos, transgenics, and cloning. This report has been updated at least once a year for the last few years.

Directorate-General Communication. Directorate-General for Health and Consumer Protection. European Commission. Gallup Organization. *Europeans' Attitudes Towards Animal Cloning: Analytical Report.* Brussels: European Commission, 2008.

This report summarizes the views of more than 25,000 individual citizens over the age of 15 in 27 member states of the European Union interviewed between July 3 and 7, 2008. Survey questions dealt with topics such as an understanding of the nature of animal cloning by participants, views about cloning for the purposes of food production, trusted sources of information about cloning, purchasing food produced by cloning, and attitudes toward the labeling of foods produced by cloning.

Javitt, Gail H., Kristen M Suthers, and Kathy Hudson. *Cloning: A Policy Analysis.* **Washington, D.C.: Genetics and Public Policy Center. Johns Hopkins University, 2005.**

The authors of this report point out that scientific progress in the field of cloning is moving more rapidly than public understanding of the procedure, leading to problems in developing appropriate legal and regulatory instruments for dealing with the technology. This report summarizes the result of surveys with 4,834 Americans about the technology and ethical implications of cloning, as well as a very well written explanation of the process of cloning, a review of arguments for and against human cloning, an analysis of federal oversight with respect to human cloning, state laws on the subject, and international policies with regard to human cloning.

The President's Council on Bioethics. *Human Cloning and Human Dignity: An Ethical Inquiry.* **New York: Public Affairs, 2002.**

This report of the President's Council on Bioethics first reviews the technology of cloning and then assesses the ethical implications of both therapeutic and reproductive cloning. The authors conclude their report with a number of recommendations for public policy on cloning.

Internet Sources

"Cloning." Stanford Encyclopedia of Philosophy. URL: http:// plato.stanford.edu/entries/cloning/. Accessed on March 16, 2009.

This excellent and comprehensive discussion of cloning includes separate sections on cloning for research therapy, human reproductive cloning, and religious perspectives. It also provides links to other print and electronic resources on the subject.

Genetic Science Learning Center. University of Utah. "Cloning." URL: http://learn.genetics.utah.edu/content/tech/cloning/. Accessed on March 16, 2009.

This Web site provides a clear understanding of the process of cloning and offers an interactive exercise that permits one to go through the steps in making a clone. The Web site also provides links to a number of other very interesting and useful Web pages that deal with topics such as reasons for cloning, cloning myths, the risks of cloning, and issues related to cloning.

Human Genome Project Information. "Cloning." URL: http:// www.ornl.gov/sci/techresources/Human_Genome/elsi/cloning .shtml. Accessed on March 16, 2009.

This Web site provides a comprehensive, unbiased source of information on all aspects of cloning, including a description of cloning procedures, a list of animals that have been cloned, some uses to which cloning can be put, possible risks from cloning, and issues over the cloning of human beings.

Medline Plus. "Cloning." URL: http://www.nlm.nih.gov/ medlineplus/cloning.html. Accessed on March 16, 2009.

Operated by the National Library of Medicine and the National Institutes of Health, this Web site offers a general introduction to the subject of cloning. Sections of the site deal with a basic overview of the technology, recent research in the field, organizations interested in cloning, and additional resources on the topic.

World Health Organization. "A Dozen Questions (And Answers) on Human Cloning." URL: http://www.who.int/ ethics/topics/cloning/en/. Accessed on March 16, 2009.

The World Health Organization poses a dozen fundamental questions about cloning and provides authoritative answers for each. The questions deal with issues such as what a clone is, whether cloning can occur naturally or not, why scientists do experiments on cloning, how reproductive and therapeutic cloning differ from each and what the purpose of each is, and what the ethical issues surrounding cloning are. The Web site also provides links to a number of policy statements on cloning and some useful resources on the topic.

Glossary

allele   One member of a pair or series of different forms of a gene. An allele is a specific sequence of nucleotides on a DNA molecule.

amino acid   A biochemical compound containing the amino group (-NH$_2$) and one or more carboxyl groups (-COOH). Proteins are made of many (dozens, hundreds, or thousands) amino acids joined to each other.

amplification   The process by which the number of copies of DNA is increased.

analyte   A substance that is measured in a laboratory test, such as a specific hormone or enzyme.

bacteriophage   *See* phage

base pair (bp)   Two nitrogen bases (either adenine and thymine or cytosine and guanine) joined by hydrogen bonds in a DNA molecule.

base sequence   The order in which nitrogen bases occur along a DNA molecule. Base sequence determines the protein formed by the DNA molecule.

bioballistics   A method for inserting genes into a host cell in which then metal slivers are coated with genes and fired into the cell by some type of propulsive device.

candidate gene   A gene thought to be responsible for some specific disease.

chemical poration   A method for inserting genes into a host cell in which the cell is treated with some chemical that will produce tiny holes in the cell walls, allowing genes to be inserted into the cell more easily.

chimera   An organism produced by the combination of two distinctly different organisms, as in the insertion of human gene into a mouse or the cross-breeding of a sheep and a goat.

chromosome   A thread-like structure found in the nucleus of a cell that consists of DNA and a variety of structural proteins.

clone   An organism that is identical in its DNA composition to some other organism, produced by some form of asexual reproduction.

codon   A sequence of three adjacent nucleotides on a strand of DNA or RNA that specifies the genetic code for the synthesis of some specific amino acid.

complementary DNA (cDNA)   A single strand of DNA synthesized in the laboratory from an RNA molecule using the enzyme reverse transcriptase.

confidentiality   In the field of genetic testing, the principle that information obtained from a genetic test cannot be provided to any other person without permission of the person being tested.

deoxyribose   A pentose (five-carbon sugar) that is one of the three major components of DNA. Deoxyribose has one hydroxyl (-OH) group fewer than ribose.

deoxyribonucleic acid   *See* DNA

DNA   A biochemical molecule consisting of two strands wrapped around each other in a helical formation in which the individuals strands consist of alternating units of deoxyribose sugar and phosphate groups held together by pairs of nitrogen bases.

DNA probe   *See* probe

DNA replication   The process by which a DNA molecule unwinds and makes an exact copy of itself.

DNA sequence   The order in which nitrogen bases occur in a DNA molecule.

DNA sequencing   Any set of procedures by which the exact order of nitrogen bases in a molecule of DNA is determined.

double helix   The geometric shape of a DNA molecule, consisting of two strands of polynucleotides wrapped around each other in a "spiral staircase" formation.

edible vaccine   A vaccine that consists of some edible food into which has been inserted one or more genes for an antigen.

electrophoresis   A fundamental laboratory procedure by which molecules are separated from each other on the basis of the rate at which they migrate in an electrical field.

electroporation   A method for inserting genes into a host cell by treating the cell with an electric shock to produce tiny holes in the cell wall, allowing genes to be inserted more easily.

enzyme   A protein that acts as a catalyst, that is, that increases the rate of a chemical reaction in the body.

eugenics   A philosophical system and field of study devoted to the belief that a population of organisms can be improved by selective breeding.

exogenous DNA   DNA that has been introduced into an organism.

exon   A segment of a DNA molecule that codes for a gene.

expressed sequence tag (EST)   A portion of a cDNA molecule that can be used to identify a gene.

gene   The unit of heredity. A gene is a segment of DNA responsible for the synthesis of messenger RNA in the production of a protein.

gene amplification   The process by which numerous copies of a segment of DNA are made. Gene amplification can occur naturally or as the result of experimental procedures.

gene gun   A device for firing genes into a host cell.

gene mapping   The process by which the location and linear separation of genes on a chromosome are determined.

gene therapy   Any process by which scientists attempt to cure or ameliorate the effects of a genetic disease by correcting, replacing, or supplementing incorrectly functioning or nonfunctioning genes in an organism.

gene transfer   The process by which a gene is inserted into an organism. Gene transfer is conducted for a number of different reasons and by using a variety of technologies.

genetic code   The "instructions" carried in a DNA molecule that tells that molecule what protein to make. The genetic code consists of various combinations of three nitrogen bases.

genetics counseling   The process by which information is provided to individuals about the possible risks of genetic disorders, with discussions of alternative actions that may be available. Genetics counseling is often a critical aspect of any genetic testing or screening program.

genetics discrimination   Actions taken against an individual or a group based solely or primarily on the basis of genetic characteristics.

genetic predisposition   An individual's susceptibility to a disease because of some error in his or her genetic characteristics. Genetic predisposition does not mean one necessarily develops that disease; only that he or she has the genetic characteristics for it.

genetic screening   Genetic testing on a group of people to determine their susceptibility to a particular disease.

genetic testing   The analysis of an individual's genetic characteristics to determine the likelihood that he or she may develop a disease or to confirm a diagnosis of such a disease.

genetically engineered food   *See* genetically modified food

genetically modified food A food or food ingredient consisting of or containing one or more genetically modified organisms or the product(s) of such organisms.

genome The completed genetic composition of an organism, often expressed in terms of the nitrogen base pairs that occur in the organism's DNA.

genomics The study of genes and their function.

human gene therapy *See* gene therapy

in vitro A procedure carried out in the laboratory as opposed to within a living organism (from the Latin expression for "within glass").

in vivo A procedure that takes place within a living organism, as opposed to in a laboratory setting (from the Latin expression for "within life").

informed consent The principle that one agrees to take part in some activity with the proviso that he or she completely understands the risks and benefits of that activity.

insertion In genetics, a type of mutation that occurs when some foreign material (one or more base pairs) is inserted into the DNA sequence of a gene, often disrupting that gene and the function it normally has.

intron A DNA sequence that separates the coding parts of a gene. Introns are transcribed into messenger RNA, but are cut out before synthesis of a protein from the mRNA.

junk DNA DNA that has no coding function.

knockout The process of deactivating a gene, usually for the purpose of studying the properties and function of that gene.

laser poration A method for inserting genes into a host cell by treating the cell with a laser beam, which produces tiny holes in the cell wall, making it easier to insert genes into the cell.

ligase An enzyme that catalyzes the formation of hydrogen bonds between two DNA or RNA fragments.

messenger RNA *See* mRNA

microinjection A process for introducing DNA into a host cell using a fine-tipped pipet.

mitochondrial DNA (mDNA) DNA found in the mitochondria in the cytoplasm of a cell.

mRNA A form of RNA that transmits information stored in a DNA molecule to ribosomes, where it is used in the synthesis of proteins.

nitrogen base A biochemical compound related to one of two basic compounds, purine and pyrimidine, found in DNA and RNA. Adenine, cytosine, guanine, and thymine are the nitrogen bases in DNA, while adenine, cytosine, guanine, and uracil occur in RNA.

nuclear transfer   The process by which the cell of a nucleus is removed from one cell and transferred to a second cell, whose own nucleus has already been removed.

nucleic acid   A family of organic compounds that consist of three basic parts: a phosphate group, a sugar (ribose or deoxyribose), and a small number of nitrogen bases.

nucleoside   A subunit of a nucleic acid consisting of one nitrogen base and one sugar (ribose or deoxyribose) attached to each other.

nucleotide   The basic unit of a DNA molecule, consisting of a phosphate group, the sugar deoxyribose, and one of four nitrogen bases: adenine, cytosine, guanine, or thymine.

penetrance   The tendency of a particular genotype to be expressed as a particular phenotype. If the genotype always produces the characteristic phenotype, the genotype is said to be completely penetrant. If it is expressed only some of the time, the genotype is said to be incompletely penetrant.

phage   Shorthand for *bacteriophage*, a virus that attacks bacterial cells, takes control of the cell's system of reproduction, and makes copies of itself.

pharmacogenics   The study of the way in which a person's genetic makeup affects his or her reaction to drugs.

pharming   The use of DNA technology to engineer a plant or animal in order to make a commercially useful product.

plasmid   An extra-chromosomal double-stranded circular piece of DNA found in bacteria and protozoa.

polymorphism (genetic)   A frequently occurring variation in the nucleotide sequence in a DNA molecule. Genetic polymorphisms are responsible for the production of polymorphic proteins.

positive predictor value (PPV)   The likelihood that a person will develop a disease for which he or she is being tested.

precautionary principle   The concept that regulatory agencies may be justified in taking regulatory actions even when some scientific uncertainty exists as to the possible risks and consequences of some given practice.

probe   A single-stranded DNA or RNA molecule prepared synthetically, with a specific base sequence, used to detect a DNA or RNA segment with the complementary base pattern. The probe also carries some time of label, such as a radioactive isotope, that allows its location to be tracked.

proteomics   The study of the full set of proteins produced by some given genome.

recombinant DNA (rDNA)   A synthetic DNA molecule produced by the insertion of some foreign DNA segment into a host molecule.

restriction endonuclease   *See* restriction enzyme

restriction enzyme   An enzyme that recognizes specific base segments in a DNA molecule and then cuts those segments at specific positions.

ribonucleic acid   *See* RNA

RNA   A biochemical compound that contains the sugar ribose, phosphate groups, and four nitrogen bases: adenosine, cytosine, guanine, and uracil. Various forms of RNA exist in cells, each form with a specific function.

somatic cell gene therapy   The insertion of genetic material into a cell for some therapeutic purpose. Since the cell is somatic, the gene modification cannot be passed to offspring.

Southern blot   A common laboratory procedure used to determine the base sequence in a DNA fragment.

traceability tag   A piece of DNA added to a genetically modified food product that has no effect on human health, the environment, or the organism into which it is inserted, but that provides a recognizable "address" of the company that made the product.

transcription   The synthesis of a messenger RNA molecule from the genetic code (base sequence) stored in a DNA molecule.

transfection   The introduction of DNA into a host cell.

transgenic organism   An organism formed by the insertion of some foreign DNA into its genome.

translation   The process by which the genetic information (base sequence) stored in a messenger RNA molecule is used to synthesize a protein on a ribosome.

vector   Any material, such as a virus or plasmid, used to introduce DNA into a host cell.

Index

About the Author

David E. Newton holds an associate's degree in science from Grand Rapids (Michigan) Junior College, a B.A. in chemistry (with high distinction) and an M.A. in education from the University of Michigan, and an Ed.D. in science education from Harvard University. He is the author of more than 400 textbooks, encyclopedias, resource books, research manuals, laboratory manuals, trade books, and other educational materials. He taught mathematics, chemistry, and physical science in Grand Rapids, Michigan, for 13 years; was professor of chemistry and physics at Salem State College in Massachusetts for 15 years; and was adjunct professor in the College of Professional Studies at the University of San Francisco for 10 years. Previous books for ABC-CLIO include *Global Warming: A Reference Handbook* (1993), *Gay and Lesbian Rights: A Reference Handbook* (1994, 2010), *The Ozone Dilemma: A Reference Handbook* (1995), *Violence and the Media: A Reference Handbook* (1996), *Environmental Justice: A Reference Handbook* (1996, 2009), *Encyclopedia of Cryptology* (1997), *Social Issues in Science and Technology: An Encyclopedia* (1999), and *Sexual Health: A Reference Handbook* (2010). Books for other publishers include *Recent Advances and Issues in Physics* (2000), *Sick!* (4 volumes; 2000), *Science, Technology, and Society: The Impact of Science in the 19th Century* (2 volumes; 2001), *Encyclopedia of Fire* (2002), *Recent Advances and Issues in Molecular Nanotechnology* (2002), *Encyclopedia of Water* (2003), *Encyclopedia of Air* (2004), *The New Chemistry* (6 volumes; 2007), *Nuclear Power* (2005), *Stem Cell Research* (2006), *Latinos in Science, Math, and Professions* (2007), and *DNA Evidence and Forensic Science* (2008).